THEORETICAL VIRTUES IN SCIENCE

What are the features of a good scientific theory? Samuel Schindler's book revisits this classical question in philosophy of science and develops new answers to it. Theoretical virtues matter not only for choosing theories 'to work with', but also for what we are justified in believing: only if the theories we possess are good ones (qua virtues) can we be confident that our theories' claims about nature are actually correct. Recent debates have focused rather narrowly on a theory's capacity to predict new phenomena successfully, but Schindler argues that the justification for this focus is thin. He discusses several other theory properties – such as testability, accuracy and consistency – and highlights the importance of simplicity and coherence. Using detailed historical case studies and careful philosophical analysis, Schindler challenges the received view of theoretical virtues and advances arguments for the view that science uncovers reality through theory.

SAMUEL SCHINDLER is Associate Professor at the Centre for Science Studies at Aarhus University in Denmark. He has published on methodological issues in science, discovery, the realism debate, and explanation in journals such as *The British Journal in Philosophy of Science*, *Synthese*, and *Studies in the History and Philosophy of Science*. He has received two major external grants from national research foundations in Germany and Denmark.

THEORETICAL VIRTUES IN SCIENCE

Uncovering Reality through Theory

SAMUEL SCHINDLER

Aarhus University, Denmark

CAMBRIDGE
UNIVERSITY PRESS

CAMBRIDGE
UNIVERSITY PRESS

University Printing House, Cambridge CB2 8BS, United Kingdom

One Liberty Plaza, 20th Floor, New York, NY 10006, USA

477 Williamstown Road, Port Melbourne, VIC 3207, Australia

314-321, 3rd Floor, Plot 3, Splendor Forum, Jasola District Centre, New Delhi - 110025, India

79 Anson Road, #06-04/06, Singapore 079906

Cambridge University Press is part of the University of Cambridge.

It furthers the University's mission by disseminating knowledge in the pursuit of education, learning and research at the highest international levels of excellence.

www.cambridge.org
Information on this title: www.cambridge.org/9781108435031
DOI: 10.1017/9781108381352

© Samuel Schindler 2018

First published 2018
First paperback edition 2020

A catalogue record for this publication is available from the British Library

Library of Congress Cataloging in Publication data
NAMES: Schindler, Samuel, 1980– author.
TITLE: Theoretical virtues in science : uncovering reality through theory / Samuel Schindler (Aarhus Universitet, Denmark).
DESCRIPTION: Cambridge : Cambridge University Press, 2018. | Includes bibliographical references and index.
IDENTIFIERS: LCCN 2017053780 | ISBN 9781108422260
SUBJECTS: LCSH: Science – Methodology. | Science – Philosophy.
CLASSIFICATION: LCC Q175 .S351724 2018 | DDC 501–dc23
LC record available at https://lccn.loc.gov/2017053780

ISBN 978-1-108-42226-0 Hardback
ISBN 978-1-108-43503-1 Paperback

For Flori

Contents

Figures

ix

Acknowledgements

This book is the outcome of several years of research. Everyone working in the field – or any academic field for that matter – knows that the process of bringing one's thoughts to fruition is bittersweet, and the disappointments and failures numerous. I am all the more excited, therefore, that this book has materialized.

The list of people who have helped me in the process is long. I owe tremendous thanks to Matteo Morganti for his generosity with his time for my concerns and for the sharp observations he made throughout the book. Likewise, I thank Peter Vickers for his detailed and well-taken comments in Chapters 1, 2, and 4. Kareem Khalifa, to whom I am indebted for invaluable advice from our time as fellows at the Pittsburgh Center for Philosophy of Science, kindly suggested to me a catchy name for the realist position I'm defending here. I also thank two anonymous readers for Cambridge University Press, who provided very detailed and helpful comments, and Hilary Gaskin for being such a reliable editor. Most importantly, perhaps, it was she who encouraged me to write a book in the first place.

Thomas S. Kuhn has been one of my greatest intellectual influences. His *Structure of Scientific Revolutions* first sparked my interest in the history and philosophy of science when I was still in high school. I've been fascinated ever since by the creativity of his thought and method. Although Kuhn is sometimes treated with contempt these days by both historians and philosophers, I hope to show in Chapters 3 and 7 that his work can still offer inspiration for valuable insights.

There is a long list of people who have commented on materials included in this book in one form or another and to whom I owe thanks (in alphabetical order): Allan Franklin, Andy Pickering, Anjan Chakravartty, Asbjørn Steglich-Petersen, Asger Steffensen, Attila Tanyi, Bert Leuridan, Brian Hepburn, Carl Hoefer, David Harker, Don Howard, Eric Barnes, Eric Scerri, Franz Huber, Gabriel Hendersen, Giora Hon, Hanne

Andersen, Hasok Chang, Heather Douglas, Helge Kragh, Henk de Regt, Henrik Sørensen, Ioannis Votsis, Jacob Busch, J. C. Bjerring, Johanna Seibt, Jon Norton, Kent Staley, Ludwig Fahrbach, Mads Goddiksen, Marcel Weber, Mattias Skipper, Michael Schmitz, Michela Massimi, P. D. Magnus, Sara Green, Slobodan Perovic, Sorin Bangu, Stathis Psillos, Theodore Arabatzis, Thomas Reydon, Tim Lyons, Wolfgang Freitag, and Wolfgang Spohn. Lois Tardío was the copy editor for this book, Lori Nash kindly produced the index for this book and Hreinn Gudlaugsson drew some of the figures. I thank them for their good work. I should furthermore thank a whole army of anonymous referees for various journals who have caused me much despair but, in retrospect, helped me significantly improve my work. To those I have failed to mention here, I apologize.

Although this book contains no material from my dissertation, I had already started thinking about some of the topics of this book back then. I'm most grateful for the patient guidance I received from Steven French, my PhD supervisor at the University of Leeds, in my first attempts to produce scholarly work. My PhD advisor Chris Timpson and mentor Adrian Wilson were very good at challenging and supporting me, respectively. Going back even further, the philosophy of science courses I attended as an undergraduate at the University of Osnabrück and McGill University in Montreal with Martin Lang, Mario Bunge, and Brendan Gillon were instrumental in affirming my plan to study the history and philosophy of science. I'm forever grateful to Nikos Green, Julia Thiesbonenkamp-Maag, Ingo Jung, and in particular Johannes Rüter for being such good friends during this time.

My parents Karla and Jürgen, my siblings, Aaron, Carlina, and Jonathan I thank for their support and belief in me throughout all these years. My biggest thanks belong to my wonderful wife Florence So for bringing such joy into my life, for her invaluable advice and support, and for just making me a better person. I dedicate this book to her.

I acknowledge permission by the publishers of the following journals to use portions of these previously published papers in this book: 'The Kuhnian Mode of HPS', *Synthese*, December 2013, 190 (18), 4137–4154; 'Novelty, Coherence, and Mendeleev's Periodic Table', *Studies in History and Philosophy of Science*, March 2014, Vol. 45, 62–69; 'A Matter of Kuhnian Theory-Choice? The GSW Model and the Neutral Current', *Perspectives on Science*, 2014, 22 (4), 491–522; 'Theoretical Fertility McMullin-Style', *European Journal for the*

Philosophy of Science, January 2017, 7 (1), 151–173, 'Kuhnian Theory Choice, Virtue Convergence, and Base Rates', *Studies in the History and the Philosophy of Science*, August 2017, Vol. 64, 30–37; and 'A Coherentist Conception of Ad Hoc Hypotheses', *Studies in the History and the Philosophy of Science*, February 2018, Vol. 67, 54–64.

Introduction

Philosophy of science is coming to resemble science: ever more specialized. Philosophers of physics deliberate the correct interpretation of quantum mechanics, whether spacetime is substantival or relational, and whether the time asymmetry of our world can be reconciled with the time symmetry of the laws of physics. Philosophers of biology grapple with determining the meaning of concepts such as gene, evolution, and fitness. Philosophers of chemistry ponder whether chemistry is really reducible to physics, and philosophers of climate modelling debate whether or not computer simulations can be trusted as much as experiments. While all these questions and debates are highly interesting and important, this *particularist trend* in the philosophy of science, as one might call it, has unfortunately gone to the expense of 'larger picture' questions about science: What is science? Is there a scientific method? How can we distinguish between science and pseudoscience? What constitutes a good scientific theory? Can science discover truths about the world? It bears some irony that today's philosophers of *science* have much to say about physics, biology, or chemistry but little about science. This book returns to one of those more ambitious philosophical questions about science, namely the question of what features characterize good scientific theories: what are *theoretical virtues* in science?

While this question clearly has a normative aspect, which I explore in this book, a successful answer will also have to take into account which properties of a theory scientists *actually* value when they decide to adopt a theory. I seek to establish this link to scientific practice primarily through various historical case studies in the venerable tradition of Duhem, Kuhn, Feyerabend, Lakatos, and others. Although any empirical or descriptive effort such as this one naturally comes with a certain inductive risk, it is an all-too-common fallacy to believe that due to the complexity and diversity we find in the sciences, any attempt to say something more general about science must be futile. Naturally, any empirical study of science will have to be carried out in one of the disciplines of science – such as physics, chemistry, and biology – as there can't be an empirical study of science *per se*. Crucially, though, true statements about theoretical virtues in any of those disciplines will automatically be true statements about theoretical virtues in science; to deny this would be to deny that any of those disciplines is a science, which would obviously be absurd. Such statements, of course, need not be true about *all* sciences or, even more implausibly, exhaust all there is to be said about theoretical virtues in science, but that need not be one's ambition. It's not mine, in any case.

Apart from attempting to answer the question What is a good scientific theory? this book also seeks to address the question Can our best scientific theories help us discover what is real? In fact, these two questions are intertwined: a good answer to the latter presupposes a good answer to the former, since we can judge whether a theory is likely to be true, if at all, only via the (internal and relational) properties of the theory. On the basis of my answer to the first question, I will argue that the 'mainstream' defence of realism, which is built on the idea that a theory's successful prediction of novel phenomena (or 'novel success' for short) is a theory's best evidence, is not justified. Instead, I think that realism can and should be defended along lines that are more in tune with the way in which theoretical properties are actually valued by scientists.

I will advance one 'central' and three 'auxiliary' arguments for realism. My central argument for realism is that a very virtuous theory – i.e., a theory possessing all of the standard virtues – is likely to be true. My three auxiliary arguments are as follows: (i) contrary to the standard view, there is an epistemic – i.e., knowledge-related – rationale for a controversial theoretical virtue, which is usually thought to be merely pragmatic; (ii) non–ad hocness is a sign of truth; and (iii) actual theory-choice decisions by scientists force us to accept that some theoretical virtues are indeed at least weakly epistemic. I will refer to the central and the three auxiliary arguments as my four *virtuous arguments for realism* and to the resulting position as 'virtuous realism'. Although all of these arguments are fairly independent, the three auxiliary arguments, as we shall see, support my central argument for realism.

I proceed in the following manner. In Chapter 1 of this book, I first introduce what many consider to be the standard theoretical virtues and the so-called explanatory defence of scientific realism. The success of the latter, it is well known, depends on whether theoretical virtues are truth-conducive. I will discuss this question with a particular focus on simplicity – perhaps the most controversial theoretical virtue. Many philosophers hold that simplicity cannot be truth-conducive because we would have to presuppose that the world is simple. But this is wrong-headed. Simplicity *can* be a reasonable epistemic concern without this presupposition. On the basis of what I call the *evidential-explanatory rationale* for simplicity, I will advance *my first virtuous argument for realism*. I call it the *argument from simplicity*.

In Chapter 2, I discuss an antirealist challenge that has been at the forefront of the realism debate in recent years, namely the so-called pessimistic meta-induction, and its cousin, the problem of unconceived alternatives, both of which appeal to the historical record of empirically successful but false theories. My main focus in this chapter, though, is on a challenge that has shed doubt on the very possibility of resolving the realism debate through any historical examples. The charge is that the 'base rate' of true theories, which is needed to compute the probability of a theory being true given its success, has been neglected and that we have no way of accessing it anyway. I take on this challenge and argue on the basis of the Kuhnian framework of theory choice and an important epistemological insight that a very virtuous theory is likely to be true when it is embraced by numerous scientists – even when the base rate is diminishingly small. I call this the *no-virtue-coincidence-argument*. It is *my second, and central, virtuous argument for realism*.

In order to fend off pessimistic meta-induction, realists have sought refuge in what has come to be known as the *divide et impera* move. That is, realists have restricted their commitments to those parts of theories which are responsible for empirical success and, more specifically, for *novel* empirical success, i.e., the successful prediction of novel phenomena. But is such a restriction warranted? Here I am doubtful. Such a restriction is only justifiable if a case can be made for the view, also known as *predictivism*, that novel success is better empirical evidence than non-novel success. However, the rationales that have been proposed to justify this view do not hold water, or so I shall argue in detail in Chapter 3. Novel success can be viewed as a form of a theory's fertility. There is another form of theoretical fertility which has received much less attention from philosophers. I shall explore this other kind of fertility, and arguments for realism based on it, in Chapter 4. My verdict here will also be negative, unfortunately.

The allegedly special epistemic status of both kinds of theoretical fertility has been motivated via an avoidance of ad hoc hypotheses. But what does ad hocness mean in the first place? This question will be the main focus of Chapter 5. The answer to this question seems intuitively clear: ad hoc hypotheses are those that are devised to save a theory from refutation. Such answer, however, tells us only about what motivates the introduction of such hypotheses, not about what is epistemically amiss with them. The latter we do need to know in order to understand why ad hocness is generally viewed as affecting a theory's empirical support in a negative way. My proposal is that ad hoc hypotheses are hypotheses that do not cohere either with the theories they amend or with the available background theories. Coherence, in turn, I believe can be understood as the provision of *theoretical* reasons for belief. This will form the basis for *my third virtuous argument for realism*. I call it the *argument from coherence*.

My *fourth virtuous argument for realism*, to be presented in Chapter 6, is based on the observation that, as a matter of historical fact, theoretical virtues have functioned as 'confidence boosters': scientists adopted (and did not just pursue) virtuous theories that resulted in ground-breaking discoveries despite the fact that these theories were contradicted by some of the available evidence. My argument for realism takes the form of a dilemma for the antirealist: either the scientists in question made ground-breaking discoveries *despite* making utterly irrational and methodologically wrong choices, or theoretical virtues are epistemic. I argue that we should try to avoid the first horn of the dilemma. This is my *argument from choice*.

In Chapter 7, I reflect on my chosen method of a historically informed philosophy of science. I argue that there are two fruitful roles for history and philosophy of science: (i) rational reconstruction of scientific practice and (ii) clarification of concepts used by scientists. Although rational reconstruction has a bad reputation in many quarters of philosophy and especially in history, I argue that there is a perfectly respectable way of doing it. Concept clarification, I argue, deserves more attention than it has received hitherto in the history and philosophy of science. Chapter 8 contains my conclusion. I end the book with an epilogue on the demarcation problem – that is, the problem of distinguishing science from non-science.

CHAPTER I

Theoretical Virtues, Truth and the Argument from Simplicity

What are the characteristics of a theory which scientists value and which guide them in their choice to adopt one theory or another? In other words, what are the *virtues* of a scientific theory? Do theoretical virtues give us good reason for belief that the theory in question is likely to be true? Our first concern in this chapter will be to develop a better understanding of theoretical virtues (Section 1.1). After introducing the so-called explanatory defence of realism and its criticisms in Section 1.1, we will then see how the realism debate turns on theoretical virtues (Section 1.3). I will argue that common reasons for rejecting the epistemicity of the perhaps the most controversial virtue, namely simplicity, are mistaken. Indeed, in my first virtuous argument for realism, I will conclude that there are grounds for believing a theory on the basis of its simplicity (Section 1.4).

1.1 Theoretical Virtues

T. S. Kuhn, in one of his later works following his highly influential *Structure of Scientific Revolutions*, listed altogether five of those theoretical virtues, which he considered to be standard (though not exhaustive) and which he thought to be '*the* shared basis for theory choice': empirical accuracy, consistency, scope, simplicity, and fruitfulness (Kuhn 1977a, 322).[1] A property that Kuhn did not list explicitly as a separate virtue, but one that is widely regarded as a minimal condition for a good scientific theory, is of course testability. Furthermore, it is widely agreed that a theory ought not to be ad hoc. There are other virtues which one might mention (e.g., symmetry, visualizability, and conservativeness),[2]

---

[1] Theoretical virtues are also sometimes denoted as 'values'. In fact, Kuhn himself suggested that label. I prefer 'virtues', because values have ethical connotations. Recently there has been a debate about the virtues of the *scientists* making theory choice (Stump 2007; Ivanova 2010). My discussion, instead, focuses on the virtues of *theories*.

[2] See, e.g., Quine and Ullian (1970), McAllister (1999), de Regt (2001).

but I will restrict my discussion to Kuhn's five standard virtues, testability, and ad hocness.

1.1.1   Testability and Non–Ad Hocness

No other criterion of a good scientific theory is as widely recognized as the falsifiability or testability of a theory – not only within the philosophy of science, but also way beyond it. Popper (1959a) famously promoted this criterion in his criticism of psychoanalysis. Basically, a theory or hypothesis must be falsifiable for it to be scientific; i.e., it must make claims about the world which can be shown to be false. This is clearly the case for Popper's favourite example of a scientific theory, namely Einstein's theory of general relativity, which, amongst other things, predicts light-bending around massive bodies in spacetime. Popper argued that psychoanalysis, in contrast, does not 'stick out its head'. No matter what the world is like, Popper claims, it can always be accommodated by psychoanalysis, as for example the Adlerian inferiority complex can accommodate both a man sacrificing his life in his attempt to save a child from drowning (in order to prove to himself how brave he is) and a man pushing a child into the water with the intent of drowning the child (in order to prove to himself that he is brave enough to commit a crime) (Popper 1963/1978, 35).[3] At the same time, Popper realized that

> no conclusive disproof of a theory can ever be produced; for it is always possible to say that the experimental results are not reliable, or that the discrepancies which are asserted to exist between the experimental results and the theory are only apparent and that they will disappear with the advance of our understanding. (50)

Popper therefore understood that theories are never tested in isolation, but always together with a host of auxiliary assumptions (this thesis is known as the Duhem–Quine thesis). Accordingly, negative test results never unequivocally indicate the falsity of the theory in question. Since, as Popper himself put it, a 'purely logical analysis of science' is not up to the task, it must be complemented by a study of *methods* – i.e., a study of 'our manner of dealing with scientific systems: by what we do with them and what we do

---

[3] Grünbaum (1979) criticized Popper for using this 'contrived' example without actually engaging with psychoanalysis. Contra Popper, Grünbaum argues that psychoanalysis is falsifiable and that its proponents were willing to give it up when they were at odds with the evidence, as in the case of Freud and his theory about the causes of hysteria, which he gave up once he realized that it was contradicted by the evidence at his disposal.

to them' (50). In particular, Popper believed, a good method of science must forbid auxiliary hypotheses which are introduced in an ad hoc fashion to save a theory from refutation and which minimize the degree of falsifiability of the overall theoretical system (42, 83). Lakatos (1978) therefore even once characterized Popper's falsificationism as 'introducing new, non-justifiationist criteria for appraising scientific theories *based on anti-ad hocness*' (39; emphasis added). We shall discuss in full detail conditions for ad hocness, as proposed by Popper and others, in Chapter 5.

1.1.2 Empirical Accuracy

A theory is empirically accurate when what it says about the world is consistent with observations. Kuhn described it as 'the most nearly decisive of all the criteria' but 'never or seldom sufficient for theory choice' (Kuhn 1977a, 323). This sentiment, it is fair to say, is widely shared by philosophers, as no evidence entails any particular theory (cf. Section 1.2.4).

Kuhn distinguished between qualitative and quantitative empirical accuracy, where the former could be described as accuracy with regard to the existence of certain observable phenomena (e.g., retrogressive planetary motion, cloud chamber tracks, light-bending) and the latter as accuracy with regard to numerical outcomes of experiments or observations or simply measurements.[4] Whether qualitative or quantitative empirical accuracy, a theory's testability is clearly a precondition for it to be able to exhibit *genuine* empirical accuracy: a theory that cannot be shown to be false on the basis of empirical tests should not receive any credit for being able to accommodate some phenomena (this is Popper's point, naturally). And of course, both forms of empirical accuracy come with an inductive risk: a theory may be empirically accurate at one point in time and cease to be so at a later point in time.[5]

1.1.3 Consistency: Internal and External

Kuhn distinguished between internal and external consistency (but then went on to collapse them). A theory is internally consistent when no contradictions (of the form p and not-p; e.g., 'it rains' and 'it does not

---

[4] Bogen and Woodward (1988) seem to make a similar distinction between phenomena and data. See my works (Schindler 2007, 2011) for a discussion of the possibility of theory-ladenness on this distinction.

[5] Empirical accuracy therefore contrasts with empirical adequacy (see Section 1.2).

rain') can be derived from it. A theory is externally consistent if it does not contradict what other well-confirmed theories say about the world.

Internal consistency may be thought to be necessary for any good scientific theory. Indeed, Popper (1940) once held that 'if one were to accept contradictions, then one would have to give up any kind of scientific activity: it would mean a complete breakdown of science' (408). It is easy to see why Popper would draw such a conclusion: from a contradiction, any proposition follows.[6] There are, however, several theories in the history of science that were embraced by the relevant scientific community and have been claimed to be inconsistent. The most prominent example is probably the Bohr model of the atom, which pictured atoms as little planetary systems with electrons circling atomic nuclei. The model has been deemed inconsistent for the following reason: according to classical electrodynamics, accelerated charged particles emit a constant stream of radiation. Yet Bohr postulated that electrons would emit no energy on their orbits around the nucleus, but only when transiting from one orbit to another. Indeed, there can be no 'ground states' in classical electrodynamics. Contrary to what Bohr presumed, electrons should spiral into the nucleus. What is one to make of such examples?

Norton (1987, 2002), for instance, argues that one can extract a consistent 'subtheory' from an inconsistent theory – such as the old theory of black-body radiation – and still recover the 'results of interest'. Or one can use an inconsistent theory such as Newtonian cosmology to arrive at a 'corrected' consistent theory. In each of these cases, one can therefore make (retrospective) sense of why scientists might have used inconsistent theories. However, it remains unclear how widely such procedures are applicable.

Frisch (2005) rejects Norton's view as too conservative. With regard to classical electrodynamics, which Frisch considers in some detail, he argues that physicists accepted the postulates of the theory 'in its entirety' (namely the Maxwell equations and the Lorentz force law) despite the fact that mutually inconsistent consequences can be derived from them.[7] In contrast to what Norton has argued in the aforementioned cases, Frisch points out that there is no consistent subset of postulates in the theory from which all of its empirical consequences could be derived.

---

[6] From the contradiction p and not-p, it follows that any arbitrary proposition q introduced by disjunction to p must be true, since (p or q) and not-p entail q.

[7] More specifically, Frisch argues that the neglect of 'self-fields' in the determination of the motion of charges in an external field with the Lorentz force law is inconsistent with Maxwell's equations and the principle of energy-momentum conservation.

Frisch concludes that consistency is 'not a necessary condition' for an acceptable theory (25) and that 'our commitments *can* extend to mutually inconsistent subsets of a theory' (41; emphasis added).[8] In particular, Frisch thinks this is permissible when 'predictions based on mutually inconsistent subsets agree approximately' (41), as in the case of electrodynamics where the violation of the conservation principle by the application of the Lorentz force to moving charges will be 'negligible' when the energy of the charge is comparatively large (42).

It is questionable whether Frisch's example really represents a counterexample to the constraint of consistency. After all, as Frisch himself notes, physicists use only consistent subsets of the theory's postulates (42). That is, physicists use the Lorentz force law for determining the motion of charges in fixed external fields and Maxwell's equations to determine the external field of a given charge, but never both the Lorentz force law *and* Maxwell's equations at the same time. Of course, the entities we are dealing with (charges and fields) are the same in both of these contexts. Ideally, one would think, they would warrant the same theoretical treatment. But physical theory is imperfect. We still lack a coherent (and fully consistent) theory of all four fundamental forces of nature, although the general theory of relativity and the standard model very successfully describe gravitational forces and the other four forces of nature, respectively.

Vickers (2013b), in his book-length treatment of several allegedly inconsistent theories (including the Bohr model and classical electrodynamics), concludes that 'most of these cases are not really inconsistent in any significant sense after all' (252). Vickers points out that any judgement about whether a theory is inconsistent depends on where we draw the boundaries of the theory in question: which propositions do we include as belonging to the theory, and which ones do we exclude?[9] This is no trivial task: if we are interested in whether a certain scientific community embraced a theory – which we probably should if we want to analyse an actual scientific theory – we would somehow have to determine where they drew the line. Or perhaps we could just ask whether particularly important scientists included certain propositions in the theory. Should we hold them responsible for only explicitly derived contradictions or contradictions one

---

[8] At the same time, Frisch adds the general qualification that an inconsistent theory cannot be a candidate for 'a universal physics'; inconsistency, for him, is only acceptable in 'a limited domain of application' (43).

[9] Vickers takes the rather extreme view that this question cannot be decided in favour of one or the other answer and campaigns for altogether *eliminating* the term 'theory' from philosophical analyses.

could derive from the propositions they hold? Another important question would be whether scientists really believed the propositions in question or whether they just used them (without believing them). We should also ask ourselves whether the use of inconsistent propositions would be enough for us to deem the inconsistency 'significant', or whether we would require belief, or at least belief that the relevant propositions are plausible truth-candidates (as Vickers suggests). Given that scientists tend to be rather tight-lipped about their doxastic attitudes, however, it may not be so easy to distinguish between mere use and belief.

In sum, although there are various philosophical views on inconsistency in science, nobody would want to claim that scientists use or even believe inconsistent theories *frequently* or that it would be good for them to do so. On the contrary, by all accounts, the use of inconsistent theories (let alone doxastic commitment to them) seems rather rare in respected scientific practice. And when inconsistency is revealed, it can be a strong reason to reject the theory in question.[10] Even if it were true that scientists occasionally use inconsistent theories (whether consciously or not), internal consistency clearly is a theoretical virtue that scientists strive for.

External consistency is also a virtue: a theory should ideally not contradict other established theories. Any theory, for example, that would violate principles of energy conservation would have to be eyed with suspicion. On the other hand, it is not rare for new theories to contradict apparently established theories. For example, when Copernicus proposed his heliocentric astronomical system, a consequence of a moving earth was the so-called parallax shift of the stars as observed from the earth. Yet, at the time of Copernicus, this observation could not be made with available instruments. Copernicus thus postulated – contrary to the received view – that the stars are extremely far away from us. He was only much later proven right (Kuhn 1957). Similarly, a common objection to Copernicus's system was that it resulted in (at the time) highly implausible physics: if the earth were to move around the sun (rather than remain at rest at the centre of the universe), then surely unsupported objects on earth would not fall to the earth in a (perceived) straight line but would rather fly off opposite to the direction of the earth's motion around the sun? It took a Newton to reconcile Copernicus's innovation

---

[10] Vickers concludes that there is probably just one 'significant' internal inconsistency of the many potential inconsistencies he considers – namely, the inconsistency between Bohr's model and the adiabatic principle, which was uncovered by Wolfgang Pauli.

with a workable terrestrial physics.[11] All this goes to show that external consistency is valued by the scientific community; at the same time (and perhaps contrary to internal consistency), it cannot be a necessary condition for a good scientific theory.

1.1.4   Unification

A theory has unifying power (or broad scope) when it unifies phenomena which were previously considered distinct. Newton's unification of terrestrial and celestial mechanics, Maxwell's unification of electrical and magnetic phenomena, the standard model's unification of the electromagnetic and the weak force, the unification of sea-floor spreading and various other geological phenomena by plate tectonics, and Darwin's unified explanation of all life on earth are all cases in point.

Philosophical accounts of unification have traditionally been accounts of explanation (and understanding) – i.e., accounts that explicate scientific explanation (and understanding) in terms of unification. The first attempt to provide such an account was made by Friedman (1974).[12] Friedman's guiding idea is that we achieve unification 'by reducing the total number of independent phenomena that we have to accept as ultimate or given', whereby Friedman conceives of phenomena as 'general uniformities or patterns of behavior' (15). Accordingly, in the model Friedman proposes, the phenomena are represented as 'law-like sentences' (15). Kitcher (1976) showed that Friedman's model suffers from an internal inconsistency, the details of which need not concern us here.

Kitcher (1981) himself developed an account of explanation as unification, which, despite severe criticism (Barnes 1992; Woodward 2003; Gijsbers 2007), remains the standard account. According to Kitcher, unifying explanations are *argument patterns*, employed repeatedly, and successfully, in order to derive descriptions of a large set of phenomena. Argument patterns consist of *schematic arguments* (which in turn consist of *schematic sentences*, structured as arguments by *classifications*) and *filling instructions*. The filling instructions tell us that, for example, the schematic sentence 'For all X if X is O and A then X is P' can be

---

[11] Vickers (2013b) argues that the alleged internal consistency in the Bohr model of the atom, mentioned earlier in this chapter, is in fact more appropriately viewed as an external inconsistency of Bohr's model with electrodynamics.

[12] Friedman was very keen to stress that he sought to provide an objective account of understanding (rather than explanation). The question of whether we need an independent account of understanding is controversial (Trout 2002; de Regt 2004; de Regt and Dieks 2005; Khalifa 2012, 2013).

instantiated as 'Organisms homozygous for the sickling allele develop sickle cell anaemia' by replacing the relevant dummy letters in the schematic sentence with non-logical vocabulary. It is easy to see that the generality of the schematic arguments allows a wide range of instantiations and thereby potentially high unifying power. Crucially, the fewer the number of argument patterns used in deriving phenomena descriptions, and the more *stringent* they are,[13] the more unified – and better – the explanation. At any one time in science, Kitcher envisions there to be a particular *explanatory store* of argument patterns that maximally unifies our beliefs about the world. Thus Kitcher, like Friedman, thinks that an understanding of the world by unification is achieved by 'reduc[ing] the number of types of facts that we have to accept as ultimate (or brute)' (432). Both Friedman and Kitcher are therefore committed to the idea that the fewer the 'ultimate facts', the higher the unifying power. Another way of expressing the idea is that when we seek to unify the phenomena, we seek to provide ever more *simple* theories. The idea of unification, then, appears to depend on some notion of simplicity, i.e., simplicity in terms of number of presumed basic phenomena, the number of basic laws, the number of argument patterns, etc.[14]

Before we move on to the virtue of simplicity, we should note that unifying power should be distinguished from the *empirical* scope of the theory. A theory can conjoin many facts and therefore have broad empirical scope but little unifying power. That would be the case if the theory would give us no clue as to how the conjoined facts are interrelated at a 'deeper' level. Conversely, a theory can *theoretically* unify the phenomena and have little empirical confirmation, as for example the Glashow-Salam-Weinberg model of electroweak interactions in the early 1970s before the discovery of neutral current, which we shall consider in more detail in Chapter 6. That model managed to make a theoretical case for unifying electromagnetic and weak interactions *before* its predictions were confirmed.

---

[13] Stringency is supposed to ensure that argument patterns capture a large set of actual arguments *in a non-trivial way*. Argument patterns, for example, should not just instantiate themselves (Kitcher 1981, 518).

[14] Simplicity considerations play a similar role in the so-called *best systems analysis* of the laws of nature by David Lewis, according to which the laws of nature are individuated by those axiomatic systems that strike the best balance between strength (understood as the amount of deductive consequences of the system) and simplicity (understood as the number of non-deduced sentences, i.e., axioms of the system) (Lewis 1986). For an illuminating recent discussion, see Woodward (2014b).

1.1.5 Simplicity

The most controversial theoretical virtue is perhaps simplicity. A common criticism of viewing simplicity as a theoretical virtue is that simplicity is indeterminate for choosing one theory over another. Kuhn (1957; 1977a, 324) points out, for example, that although the Copernican system of planetary motion, as proposed by Copernicus in 1543, was simpler than the Ptolemaic system of planetary motion when it came to describing 'merely [the planetary motion's] gross qualitative features' such as planetary retrograde motion and the limited elongation of the interior planets, it was not simpler than the Ptolemaic system when used quantitatively to describe and predict the positions of the planets in the night sky at a particular time. When used for that purpose, the Copernican system had to employ just as many epicycles as the Ptolemaic system.[15] Likewise, the Keplerian version of the solar system – in which planets traverse their orbits on ellipses rather than circles (as in the Copernican system) – can be said to be less simple than the Copernican system, because whereas ellipses are described by two parameters, circles are determined by just one parameter. On the other hand, the Keplerian version does without any epicycles and is therefore simpler with regard to the number of geometrical devices it invokes.[16] These examples illustrate that there are many different senses in which one theory can be said to be simpler than another theory. Accordingly, so the worry goes, simplicity cannot be used to decide which theory one ought to adopt, let alone which theory is more likely to be true.

Similar observations and concerns apply to a 'finer' level of resolution, for example the degree of complexity of polynomials such as $y = a + bx + cx^2 + d$. As Harre (1960) points out, there are different ways in which a polynomial might be simpler than another. A polynomial with a lower magnitude of exponents might be said to be simpler than a polynomial with a higher magnitude of exponents, so that a polynomial containing as highest exponent x^2 would then be simpler than a polynomial where the highest exponent is x^5. Or one could take the 'degree' of polynomials, viz. highest sum of exponents in any of its terms, so that $y = 15x^2z + x - 3$ would be simpler than $y = 7x^2z^3 + 4x - 9$, because the highest sum of exponents of any term in the first polynomial is 3, whereas it is 5 in the second polynomial. One could also use the number of different independent variables

---

[15] See also Gingerich (1975).   [16] This example is used by McAllister (1999, 116).

as a criterion for simplicity so that a polynomial containing just x would be simpler than a polynomial containing x, y, z.[17]

When it comes to higher-level theories, an important distinction to be made is that between *syntactical* and *ontological* simplicity or parsimony. A theory that is syntactically more parsimonious than another is a theory that postulates only a few principles or laws. For example, a theory that postulates three fundamental laws of motion, as Newtonian mechanics does, is simpler than a theory that postulates five fundamental laws of motion. Special relativity, which postulates just two basic principles – namely the principle of relativity and that of invariant light speed – is, in turn, simpler than Newtonian mechanics. At the same time, though, whereas Newtonian mechanics requires just three equations for describing gravitational phenomena (one for each space dimension), Einstein's general theory of relativity requires no less than fourteen equations (cf. Weinberg 1993, 107). A theory that is ontologically more parsimonious than another postulates fewer entities. *Ockham's razor* is typically understood ontologically as the maxim that 'entities are not to be multiplied beyond necessity'.[18]

One can further distinguish qualitative and quantitative ontological parsimony, where the former is the number of *kinds* of entities postulated and the latter the number of entities of a particular kind. Whereas qualitative parsimony is widely acknowledged to be an important consideration in theory choice (e.g., the standard model with its postulation of three kinds of quarks is preferable to a (fictional) model which postulates 10 different kinds of quarks), the latter is more controversial (e.g., Lewis 1973). Nolan (1997), on the other hand, has argued that quantitative parsimony considerations played a role in the postulation of the neutrino in the 1930s. More specifically, physicists postulated one neutrino with a spin of ½ to explain the fact that the total spin of the observed emitted particles was ½ less than the total spin of the system before beta decay. Nolan points out that the evidence was consistent with the postulation of two neutrinos with ¼ spin each. Thus physicists, at least implicitly, seem to have appealed to some kind of quantitative ontological parsimony principle.[19]

---

[17] Usually, polynomials are categorized according to the second method.

[18] Ockham (also Occam) was a fourteenth-century philosopher. Whether Ockham himself embraced this formulation is in fact not so clear (cf. Thorburn 1918).

[19] Like Nolan, Baker (2003) tries to justify the preference for quantitative parsimonious hypotheses. For recent developments regarding quantitative parsimony, see Baron and Tallant (in press) and Jansson and Tallant (2016).

The fact that there are various *forms* of simplicity makes it impossible to compare theories without first specifying the form of interest. Simplicity is thus comparable to similarity: in order to say whether a zebra is more similar to a horse than to a tiger, we need to first specify whether we want to make our judgement with regard to 'equine morphology' or to the property of 'being stripy' (McAllister 1999, 118). In fact, simplicity, just like similarity, seems to be a relative matter. A theory can be simple when compared to one theory and complex when compared to another. For example, a theory postulating 61 fundamental particles (as the standard model in particle physics does) is simpler than one that postulates 99 fundamental particles, with regard to qualitative ontological parsimony. Yet that same theory would be less simple than a theory that postulates only 32 fundamental particles – again, with regard to qualitative ontological parsimony. Simplicity might also be analogous to similarity in another regard: anything is similar to anything else, at least in some respect. As Lakatos (1970b) once claimed, 'simplicity can always be defined for *any* pair of theories T_1 and T_2 in such a way that the simplicity of T_1 is greater than that of T_2' (131, fn. 106).

Does the relativity of simplicity make it somehow arbitrary and useless as an arbiter in theory choice? I don't think such pessimism is warranted. First of all, once it is decided which property of a theory should be assessed for its simplicity, it is a matter of fact whether one theory is simpler than another. Second, although *in principle* the number of forms of simplicity seem to be inexhaustible, at least when it comes to higher-level theories, the forms of simplicity that plausibly play a role in *actual* theory evaluation appear rather limited: earlier we saw that there appear to be only two broad classes of simplicity with regard to higher-level theories – namely, syntactic and ontological parsimony. Third, for simplicity to play a successful role in theory choice, it simply is not necessary that there be only one form of simplicity that all practitioners agree upon in *all contexts*. In some contexts, practitioners might agree on one form of simplicity; in others, practitioners might agree on another. A good philosophical account of theory choice should be open to such context sensitivity of simplicity as a theory choice criterion.[20] It may well be that the agreement there is in any one particular context is not uniform and that some individuals prefer some form of simplicity over another form, and other individuals still another form. But

---

[20] An account of simplicity that context-sensitive is offered by Sober (1990). Yet, in contrast to what I proposed here, Sober is a reductionist (if not even an eliminatavist) about simplicity. See Section 1.3 for further details.

as long as the majority of the relevant scientific community can agree on one form of simplicity, then simplicity will be functional as a theory choice criterion. For example, Weinberg (1993) – who notes the aforementioned higher complexity of Einstein's general theory of relativity with regard to the number of equations required for accounting for gravitational phenomena – is adamant that the simplicity of Einstein's theory of general relativity with regard to the equivalence of gravitational and inertial mass, which Newtonian physics postulated as separate entities, 'was largely responsible for the early acceptance of Einstein's theory' (107).

Why should a rational agent prefer simple theories (or, better, theories that are simpler than others) in the first place? Popper (1959a), for example, proposes that a preference for simpler theories has to do with falsifiability. In fact, Popper believes that the simplicity of a theory comes down to degrees of falsifiability: the more falsifiable a theory, the simpler it is. Since falsifiability for Popper is something science does and must strive for, simplicity is too. Popper illustrates this with a simple example: the linear equation $y = a + bx$ is intuitively simpler than the parabola $y = a + bx + cx^2$, because in order to 'falsify' the former, only three data points are needed, whereas at least four data points are required to 'falsify' the latter. Popper also says that the simpler equation has more 'empirical content' because it prohibits more facts about the world.

There is an obvious problem with Popper's notion of simplicity. Say we conjoin some falsifiable theory T_0 with an independently falsifiable hypothesis h_1 to form a new theory T_1. We have thereby increased the falsifiability. At the same time, however, it seems that we have increased the complexity of T_0. Say we conjoin further falsifiable hypotheses to T_0, so that $T_n = (((T_0 + h_1) + h_2) + h_3) + \ldots h_n$. Clearly, T_n is much more falsifiable than T_0, albeit way less simple than T_0.[21] Another criticism that besets Popper's entire account is that falsificationism cannot accommodate probabilistic statements. For example, regardless of the number of coin flips, any sequence of heads and tails is logically compatible with the hypothesis that the coin is fair. With regard to simplicity, Popper concluded that all probabilistic theories are equally and infinitely complex (Popper 1959a, 141; cf. Sober 2015, 97). But given the ubiquity of probabilistic theories in modern science, this is clearly highly unsatisfactory.[22] Yet another objection to Popper's notion of simplicity concerns his particular

---

[21] This point is made by Maxwell (2002).
[22] A well-known proposal which explicitly incorporates probabilities is the one by Jeffreys (1973): for Jeffreys, who presumes the Bayesian framework of confirmation, simpler theories have a higher prior probability than more complex theories, and therefore yield higher posterior probabilities, given the

example of linear and parabolic equations and his readiness to measure simplicity in terms of number of adjustable parameters: when we are considering specific curves in which the coefficients have been fixed, a single observation will suffice to 'falsify' both linear and parabolic curves. We have no way of telling them apart by way of Popper's simplicity criterion (Sober 2001).

In fact, much of the philosophical discussion of simplicity in science is concerned with the problem of *curve-fitting*. The literature is vast.[23] Here I want to briefly mention one very influential proposal, which links simplicity in terms of paucity of parameters to predictive power. In model selection, Forster and Sober (1994) point out that there exists a trade-off between the empirical accuracy of a model and its simplicity in terms of number of parameters: the more parameters a model possesses (i.e., the more complex the model is), the more data it can accommodate, and, vice versa, the simpler the model, the less data it will fit. But contrary to naïve empiricist intuitions, a good model mustn't fit *all* the available data at the expense of complexity, because that would considerably decrease its ability to accommodate *future* data (i.e., the more parameters a model possesses, the less 'flexible' and the less 'tolerant' it becomes with respect to data that differ from the data that were used to fix the parameters in the original model). In particular, a good model *ought not to* accommodate experimental noise. We shall discuss Forster and Sober's proposal in more detail in Section 3.2. Suffice it to say here that although Forster and Sober's argument is plausible as far as it goes, it is questionable how it could be extended beyond the context of data modelling towards higher-level theories like Einstein's theory of general relativity. This is important with regard to the realism debate, to be introduced later in this chapter, which is about theories that postulate things we cannot directly observe.

A compelling, but often overlooked, rationale for preferring simpler theories has to do with the *evidential support* of our explanations. Entities postulated by our theories gain support only by virtue of helping us explain the phenomena. If an entity is explanatorily superfluous in the sense that a theory can explain the phenomena just as well without that entity, then

---

evidence. Jeffreys's 'simplicity postulate', however, does not explain why we should think that simple theories should have higher probability in the first place. An alternative, later proposal is that there is a link between the simplicity of a theory and its likelihood. This proposal also depends on the assumption that simpler theories are ones with fewer free parameters. See my discussion of curve-fitting in what follows for more on this assumption.

[23] For a good, comprehensive overview of various measures of simplicity, in particular probabilistic measures, see Sober (2015).

the entity should not receive any support from the phenomena that the theory explains. Hence, adopting a theory that postulates explanatorily superfluous entities, rather than a theory that successfully does so with fewer entities, amounts to adopting a theory that postulates empirically unsupported entities. The same reasoning applies to syntactic parsimony. Let us refer to this rationale for the preference for simplicity in theory choice as the *evidential-explanatory rationale*.[24]

As Barnes (2000) has argued, the evidential-explanatory rationale for simplicity and non-explanatory rationales for simplicity can come apart. Consider two theories T_1 and T_2 that successfully explain the same set of data D. T_1 postulates fewer entities than T_2 and is therefore simpler. However, whereas *all* entities within T_2 are required (within T_2) to explain D, T_1 postulates an entity that is not required (within T_1) to explain D. A non-explanatory strategy would have us pick T_1, i.e., a theory with fewer entities, some of which are explanatorily superfluous. The explanatory strategy would have us pick T_2 instead, i.e., a theory that postulates only entities that are required to explain the data (354).[25]

Similarly, a theory which postulates only 32 fundamental particles, despite being simpler than a theory which postulates 61 fundamental particles (as the standard model does), would not automatically be preferable to that 'more complex' theory. According to the evidential-explanatory rationale, it would only be preferable if those 29 additional particles were not needed to explain the relevant data. In other words, explanatorily impoverished theories, which do not postulate all the entities needed to explain the phenomena, would not be preferable, even when they are simpler. The evidential-explanatory rationale for simplicity considerations in theory choice is therefore *context-dependent*.

There is another way to spell out the rationale for choosing simpler theories in terms of explanation. As we saw in Section 1.1.4, the leading

---

[24] According to Fitzpatrick (2013), this justification can be found in Mill (1867, 526). Barnes (2000) offers a probabilistic reconstruction where P(superfluous entity | theory and *evidence*) = P(superfluous entity | theory), which just says that the probability of the superfluous entity is not raised by the evidence; it has no evidential support. Barnes also points out that although we know a priori – by virtue of the Kolmogorov axioms of probability theory – that the probability of a conjunct (such as theory and superfluous entity) is always lower than any of its parts alone, this does not suffice to motivate the evidential-explanatory rationale of simplicity, since this point concerns only logical strength and not the evidence. More generally, weaker logical strength (as in a theory without a superfluous entity) per se shouldn't be a theory choice criterion, as that would amount to preferring theories on the basis of them taking lower risks of being refuted.

[25] Barnes claims (with some evidence) that most scientific episodes are best viewed in terms of the evidential-explanatory rationale rather than non-explanatory ones.

accounts of explanatory unification view unification as the explanation of a wide range of phenomena by means of a few basic assumptions/phenomena/laws/argument patterns. The 'simpler' the set of those 'bases', the better the explanatory unification. One may thus try to motivate the view that simplicity is an epistemic virtue by pointing to unification: since unification clearly is a virtue and since unification depends on simplicity, simplicity itself ought to be virtuous. Or maybe it's the other way around: is unification a virtue only because simplicity is? Usually, however, unification is taken to be less problematic than simplicity. In any case, the uncontroversial virtue of unification seems strongly intertwined with simplicity.

Overall, evidential-explanatory justifications for the use of simplicity as a *rational* theory choice criterion strike me as most promising. Whether simplicity can be considered an *epistemic* theoretical virtue – i.e., a virtue that makes a theory more likely to be true – is another issue. We shall take up this conversation again in Section 1.3.

1.1.6 Fertility

Kuhn characterized fertility as a theory's capacity to 'disclose new phenomena or previously unnoted relationships among those already known' (Kuhn 1977a, 322). These days, theoretical fertility is normally understood in terms of novel success, i.e., in terms of a theory's confirmed predictions of novel phenomena. Novel success has played a central role not only in some classical discussions about theory choice (Lakatos 1978; Worrall 1989b), but also in the realism debate: realists are normally willing to commit to the truth of only those theories which have managed to produce novel success (Psillos 1999). There are various ways in which to understand and to motivate the importance of novel success. We shall discuss them in Chapter 3, although we shall first shed more light on the role of novel success for the realism debate in Chapter 2. In Chapter 4 we shall discuss a form of fertility other than novel success.

This concludes our short overview of the current state of affairs concerning theoretical virtues. As mentioned in the Introduction, the question about what it is that makes a scientific theory a good one and the question of whether our best scientific theories can help us discover what is real are interrelated. Before we can see in more detail why that is the case, we must get a better grip on what precisely is at stake in the realism debate.

1.2   The Explanatory Defence of Realism and Its Criticisms

Contrary to what the uninitiated may think, the realism debate in the philosophy of science has nothing to do with whether or not the real world exists. Instead, it concerns our right doxastic attitudes, namely what we can or should believe. In particular, the debate has focused on things we cannot observe – *unobservables*. Whereas realists argue that our best scientific theories give us good reasons to believe in the existence of atoms, electrons, fields, genes, tectonic plates, etc., antirealists deny that.

Generally, we tend to believe that those things that we see exist: tables and chairs, the cars down at the parking lot, the trees along the alley, etc. Our senses can of course mislead us, and we can fall victim to optical illusions such as seeing an oasis after a long walk through the desert without water. But these cases are rare and generally our senses are very reliable. In contrast, there are many things which we cannot see, which we tend not to believe: witchcraft, the healing power of certain stones, homeopaths' claim that water has 'memory', Heidegger's nothingness, etc.

Antirealists take what they believe to be a healthy critical attitude to all things unobservable. Although antirealists grant that forces, quarks, electric fields, genes, black holes, and other entities postulated by our best scientific theories *may* exist, they are adamant that the evidence science provides does not allow us to infer that those entities *do* exist. This is why antirealists have described their doxastic attitudes towards theoretical entities as *agnostic* (van Fraassen 1980).

Antirealists argue that the practice and success of science can very well be described without commitment to what our best scientific theories tell us about unobservables. Although we may indeed interpret our theories literally – i.e., as referring to, e.g., electrons – doxastic commitment to the referents of the relevant terms (such as 'electron') is extraneous.[26] Instead, we only need to believe that our theories are *empirically adequate*, i.e., consistent with the phenomena in the present, in the past, and (more controversially) in the future. In other words, we only need to believe that our best scientific theories successfully 'save the phenomena'.

---

[26] The logical positivists, contrary to contemporary antirealists, sought to reduce theoretical terms such as 'electron', 'field', etc., in the language of science to observational terms. Wishing to make sense experience the objective basis of science, they had to distingusih sharply between observational and theoretical language. Modern antirealists, however, accept that all observation is theory-laden and that no such sharp line can be drawn between vocabularies. The way in which the leading antirealist, van Fraassen, has drawn the line between observable and unobservable *entities*, however, is highly controversial (cf., e.g., Churchland and Hooker 1985).

1.2.1   Unachievable Truth?

Some philosophers of science are avowed antirealists because they do not believe that science could ever achieve the truth or that we, qua our limited cognitive capacities, simply have no means for assessing whether our theories are actually true. T. S. Kuhn, echoing many laypeople's sentiments about the topic, writes this in the postscript to his *Structure of Scientific Revolutions*:

> There is, I think, no theory-independent way to reconstruct phrases like 'really there'; the notion of a match between the ontology of a theory [viz., its theoretical terms] and its 'real' counterpart in nature now seems to me illusive in principle. (Kuhn 1962/1996, 206)

Popper (1959a), too, was sceptical about truth, but for different reasons. Popper knew that there is an asymmetry between confirmation and falsification: whereas no amount of evidence can ultimately confirm a theory (because the relevant evidence can always prove us wrong), a single piece of evidence can falsify a theory – at least in principle (cf. Section 1.1.1). Thus, whereas we can *know* when a theory is false, we can *never know* when it is true. Popper initially thought it might be better to avoid talk about the truth of scientific theories (273) and also reports that he once believed truth to be a 'vague and highly metaphysical notion' (Popper 1963/1978, 231). Later Popper eased up to the notion of truth and tried to make precise the idea of the truth-content of a theory or *verisimilitude*, which he defined in terms of the logical, empirically testable consequences of a theory. He maintained that there are 'no general criteria by which we can recognize truth', only 'criteria of progress towards the truth' (226). Whether or not we could obtain truths about unobservables Popper did not comment on.

The vast majority of contemporary philosophers of science believe that the statements of scientific theories are capable of being true and that, in principle, they are true exactly when they correspond to the facts or the way the world is, just as the Tarskian correspondence theory of truth would have it. (For example, a theoretical statement such as 'there are electrons' is true exactly when there really are electrons in the world.) They are therefore no different from statements of the more pedestrian sort, such as 'it is raining', which is true exactly when it is raining. Of course, whether or not there are electrons is more difficult to assess than whether or not it is raining. We cannot directly check whether or not there are electrons; they are unobservable. Antirealists therefore recommend that we make no commitments regarding the truth-value of theoretical terms like 'electron'. Realists, on the other hand, are more confident: they typically use explanationist strategies to make their case. This shall be the topic for the remainder of this chapter.

Despite their optimism towards assessing truth claims regarding unobservables, however, realists share Kuhn's and Popper's healthy scepticism regarding our theories ever capturing the *full* truth about their respective domains. That is, no modern realist believes that we can have enough evidence to claim, justifiably, that any of our current or future best theories is true. Nevertheless, modern realists are confident that we actually succeed in constructing theories that get us closer and closer to the truth.[27] For example, the theory of relativity is closer to the truth than Newtonian mechanics or even Aristotelian physics; the belief that combustion is attributable to the chemical reaction of oxygen with other elements is closer to the truth than the belief that combustion amounts to the giving off of phlogiston, etc. In order to express the idea of improvement and progress, we are bound to appeal to truth – or so the realist has it. Antirealists of course disagree. For them, the idea of progress can be captured in purely empirical terms: we simply manage to accommodate more and more facts with our theories, i.e., we are improving the empirical strength of our theories. But how do we explain the fact that our theories are empirically successful and empirically progressive? This is where the explanationist defence for realism comes in.

1.2.2   The Local Argument for Realism: Inference to the Best Explanation

Perhaps the most straightforward realist position is to argue that the inferences made by scientists about unobservables such as electrons, electric fields, spacetime, etc., should be accepted as reliable and their conclusions as true. But what exactly are these inferences in question? Realists believe that they have the following general structure:

We make some observation O.
We surmise that hypothesis H best explains O (amongst the set of hypotheses considered).
We infer that H is likely to be true.

This kind of inference is known as the inference to the best explanation (IBE), or abduction. Here is an example:

---

[27] Philosophical accounts of truth-likeness are reviewed by Psillos (1999).

P1: Cathode rays are deflected in an electrical field.

P2: This is best explained by the cathode rays consisting of negatively charged particles, namely electrons.

C: Therefore, electrons exist.

It is easy to see that IBEs are pretty ubiquitous not only in science but also in our everyday lives; for example:

P1: I run a relay race that is officially 5 km long. My GPS watch, however, indicates that I ran 5.12 km.

P2: This is best explained by the official distance of the race being false: the actual distance covered in the race is 5.12 km, not 5 km.

C: I conclude that the official distance of the race is false.

Obviously, IBE is a non-deductive mode of inference; the conclusion does not follow by necessity from the premises, as the conclusion could be false even when the premises are false. In the previous example, alternative explanations are that I did not press my GPS button precisely at the start and the finish of my race; the official race distance is actually correct. Or it could be that my watch is malfunctioning despite my having pushed the GPS button at precisely the right time. Or it could be that the GPS system is simply not accurate enough. IBEs, therefore, involve an inductive risk in the same way that classical enumerative inductions like 'swan s1, s2, s3, etc., are all white, hence all swans are white' do: they may result in false conclusions. In contrast, we are bound to conclude that 'Clara can't drive cars' from the premises 'No women can drive cars' and 'Clara is a woman' (even though the first premise is clearly false).

Realists argue that if one accepts IBE to be a reliable (albeit fallible) mode of inference in everyday contexts, one must accept it as a reliable mode of inference also in the scientific context. As Psillos (1999) puts it, 'the point is that if abductive reasoning is ontologically committing in everyday life, then there is no reason not to be so committing in science' (212). Antirealists have resisted this idea: they either deny that IBE is a reliable mode of inference when it comes to unobservables or they shed doubt on IBE being a reliable mode of inference for anything.[28] The former move can be paraphrased as an inference to the *empirical adequacy* of the best explanation. That is, the antirealist

---

[28] Both of these moves have been attributed to van Fraassen. Van Fraassen himself has committed to the latter option (Ladyman et al. 1997).

can admit that scientists infer certain hypotheses when seeking to accommodate the phenomena but can deny that scientists are entitled to infer the truth of these hypotheses. In our earlier example, that would amount to the inference that the electron hypothesis is empirically adequate, not that electrons exist (van Fraassen 1980, 20–1). This strategy, however, can be criticized as begging the question against the realist, because the antirealist would deny that a mode of inference she accepts as generally reliable extends to unobservables *just because* it concerns unobservables. She would thus need some independent reasons for why IBE is somehow less reliable for unobservables than for observables. The other antirealist strategy of questioning IBE as a reliable mode of inference *tout court*, in the face of the ubiquity of IBE in our everyday lives, should strike one as highly unreasonable too. And yet one may ask with the antirealist what it is about the explanatory virtues appealed to in IBE – some of which we listed previously – that ought to compel us, or at least tempt us, to believe the conclusion of IBE-type arguments. In other words, one may seek to undermine IBE by questioning the epistemicity of theoretical virtues. We shall return to this critique in Section 1.3.

Antirealists have presented another argument against IBE. The argument, known as that of the 'bad lot', points out the possibility that the true hypothesis might not be amongst the ones considered. Thus, the inference to the truth of one of the considered hypotheses is not warranted (van Fraassen 1989, 142–3). If the criticism is supposed to undermine IBE as a reliable mode of inference, it clearly fails. Although it can indeed not be excluded that we are considering only a bad lot of hypotheses, IBE was never meant to *guarantee* the truth of the conclusion. IBE is an inductive inference and therefore fallible.[29] Antirealists have been willing to go even further than this. Instead of pointing to the logical possibility of the bad lot, it has been argued that it is in fact *unlikely* that the true hypothesis is amongst the ones we consider (Stanford 2006). But we will cover more on that point later on (Section 1.2.4).

---

[29] Psillos (1999), in a continuation of the work of Boyd (1983) and Lipton (1991/2004), argues that our (approximately) true background theories, which supposedly are themselves the outcome of IBEs, constrain our IBEs so that it is likely that the true hypothesis is amongst the ones we consider. I think this straightforwardly begs the question against the antirealist: the antirealist's scepticism applies to *any* scientific theory. The realist therefore cannot appeal to the truth of background theories to make her case for realism.

1.2.3   The No-Miracles Argument for Realism

Putnam once famously wrote, 'realism is the only philosophy that doesn't make the success of science a miracle' (Putnam 1979, 73). How so? The argument, also known as the 'ultimate argument for realism' or the 'no-miracles argument' (NMA) for realism, starts with the observation that some of our current best scientific theories – such as those of evolution, quantum mechanics, relativity, and plate tectonics – are tremendously empirically successful. They get a huge amount of data right and often in a very precise way: quantum electrodynamics, for example, has a predictive accuracy of 10 parts in a billion (10^{-8}). Realists like Boyd (1983) and Musgrave (1988) ask how this is to be explained if not with our theories 'latching onto the world', i.e., them getting something about the world fundamentally right, *including* unobservable entities and mechanisms postulated to underlie the phenomena. Would it not border on a miracle if those theories were as empirically successful as they are and *not* be at least approximately true? As is easy to see, the NMA is a form of IBE:

P1: The tremendous success of our current best scientific theories cries out for an explanation.

P2: If our current best scientific theories were approximately true, their tremendous success could be explained. Otherwise, the success of science would be miraculous.

C: Our current best scientific theories are likely to be approximately true.

Antirealists remain unconvinced. They have offered several arguments against the NMA.

1.2.3.1   The Circularity Charge

The first charge we want to consider is that the NMA is circular: since the NMA clearly uses IBE in order to infer that our current best scientific theories are true, and since antirealists have denied that IBE is a reliable mode of inference – at the least when it comes to unobservables – the conclusion cannot be convincing to the antirealist (Laudan 1981; Fine 1986). Realists have in response granted that the NMA is rule-circular but have denied that it is premise-circular (Psillos 1999). That is, although the NMA is indeed a version of IBE, none of its premises contains what the argument seeks to establish, namely that our current best theories are likely to be true. Thus, the NMA is not viciously circular as the argument 'The Scripture is the word of God and therefore right; the Scripture says that God exists; hence, God exists' obviously is. Nevertheless, the realist does rely on the IBE when making the NMA, so if the IBE is no acceptable mode of inference, NMA will not be convincing either. Realists generally concede that someone negatively predisposed towards IBE cannot be convinced to adopt the NMA (see Psillos 1999 and Lipton 1991/2004). Indeed, realists have pointed out that even our most secure inferences, namely deductive ones, are hard to justify when faced with the sceptic. How do you convince someone without any idea of logic about *modus ponens* (if p then q; p; therefore q) being a sound inference? You might try to explain, 'Well, *if* you are given the rule "if p then q" and you are being given p, *then* you must infer q'. But that of course presupposes that you *already* accept *modus ponens* as a sound inference.[30] What realists have therefore resigned themselves to contend with is that although they won't convince the sceptic, they can at least 'reflect on the rules we use or are disposed to use uncritically' and 'examine the extent to which and in virtue of what these rules are reliable' together with those who already accept that IBE is a reliable mode of inference (Psillos 1999, 89).

1.2.3.2   *Antirealist Alternative Explanations*

Another antirealist strategy against the NMA is to offer alternative explanations for the success of science with the purpose of undermining the claim that without realism the success of science would be an inexplicable miracle. Several such alternative explanations have been offered. I shall focus on the one that is most well known and probably most compelling.[31] According to van Fraassen,

> The success of science is not a miracle. It is not even surprising to the scientific (Darwinian) mind. For any scientific theory is born into a life of fierce competition, a jungle red in tooth and claw. Only the successful theories survive – the ones which *in fact* latched on to actual regularities in nature. (van Fraassen 1980, 40)

---

[30] Worrall (1989a) reports that 'as Lakatos used to say (only half-jokingly) there comes a point when a rationalist must get out his machine-gun to defend rationality' (384).

[31] Other antirealist explanations are offered by Fine (1986, 1991) and Stanford (2000). Fine suggests that the best explanation for the success of science is instrumental reliability rather than truth. According to Stanford, the empirical success of a theory can be explained in terms of predictive similarity to true theories. Both of these proposals have been adequately criticized by Psillos (1999) and Psillos (2001), respectively. Very briefly, Psillos argues that Fine's suggestion amounts to no more than word play, because saying that our theories are successful because they are instrumentally reliable seems to say no more than that our theories are empirically adequate. But that is what the realist thinks ought to be explained. Stanford's proposal, on the other hand, seems not very different from a realist explanation: the explanatory work of success is ultimately carried out by truth.

The thought is that it is not at all surprising that our current theories are as successful as they are: they are simply those that remained after numerous less empirically accurate theories have been weeded out on the basis of not being empirically accurate.

Van Fraassen's explanation has been criticized as shallow. Just as we can explain why a certain species survives in the struggle for existence (amongst other things, because of their genetic makeup which has caused physiological features), we can explain why scientific theories are empirically successful. Consider a funny analogous example from Lipton (1991/2004): There is a group of people, each member of which has red hair. This is explained by each of those group members belonging to the club of red-haired persons. Club membership, in a sense, is a selective mechanism and therefore an explanation for those individuals having red hair. But of course we could give a deeper explanation of why the group members have red hair by consulting their genome. Now, the antirealist may respond, Why ought we to provide a deeper explanation for the success of science? All explanations have to come to an end at some point. For example, Newton explained the planets' motion in orbits around the sun in terms of the gravitational force exerted on the planets by the sun. One could of course ask further, Why do bodies attract each other? Newton had no answer to that question. But this does not at all diminish the value of Newton's explanation. His explanation is correct as far as it goes, but we do of course have a deeper explanation of gravitational forces in terms of the geometry of spacetime. Likewise, we have a deeper explanation of the success of science in terms of the NMA, or so the realist argues.

Van Fraassen is willing to go further than this. He believes that an explanation of the success of science 'really does not matter to the goodness of the theory, nor to our understanding of the world' (van Fraassen 1980, 24). Instead, the success of science can be accepted as a 'brute fact', which is not to say that it cannot have an explanation, only that it does not need one (24). Van Fraassen draws an analogy to him meeting a friend in the supermarket by coincidence. We can explain how the coincidence occurred: both he and his friend can explain why they went to the supermarket at that point in time. But none of these explanations would somehow eliminate the fact that the two friends met by coincidence. Van Fraassen's point is that there cannot be a default demand for explanation. Sometimes we just have to accept coincidences and 'accidental correlations' – in science as well as in life in general. The realist is unlikely to be moved by this. Seeking to block a demand for explanation when a good explanation is available is as irrational as insisting that no explanation is needed for why the sun rises every day in the morning and sets every day in the evening, when we clearly possess a good explanation for this in terms of the earth's daily rotation in its orbits around the sun.

1.2.4   Underdetermination: Old and New

Another argument against the NMA derives from the so-called *under-determination of theories by evidence* (UTE) thesis, which says that no evidence determines the choice of any particular theory; any evidence is compatible with a multitude of theories accommodating the evidence. But if that is the case, then – at least in principle – theories postulating radically different theoretical entities may equally be supported by the same evidence. We would thus have insufficient reason to believe in the reality of the unobservables postulated by any of our theories.

One can distinguish between permanent and transient forms of UTE: in the former, no evidence whatsoever can tell the theories in question apart (not now and not in the future), whereas in the latter, the theories in question may be underdetermined by the evidence only within a certain time window. Traditionally, antirealists have sought to argue for the permanent form of UTE, which we shall focus on in what follows. Antirealist arguments for UTE fall into two classes: global and local arguments. The former seek to establish UTE for all theories, while local strategies explore features of particular theories to establish UTE solely for *those* theories. An example of a global strategy is suggestions made by Kukla (1996). For example, the world could behave according to our best scientific theory in the relevant domain when observed, but otherwise according to some other theory postulating radically different unobservables. Stanford (2001, 2006) dismisses such global arguments as 'Cartesian fantasies' and points out that they collapse challenges to scientific realism stemming from UTE with global scepticism. The latter may be difficult to argue against (How could we ever determine that we do not live in 'the matrix'?), but it wouldn't be something that *specifically* the scientific realist needs to worry about.

Perhaps the most famous example of a local strategy to establish UTE is Newtonian mechanics in two versions: in one, it is postulated that the universe is at rest in absolute space (as Newton himself believed), and in the other, it is postulated that it is moving with a constant velocity in absolute space (van Fraassen 1980). Observationally, these two versions of Newtonian mechanics are indistinguishable. UTE thus seems supported. And yet it is questionable whether these two versions of Newtonian mechanics are sufficient to motivate arguments against realism. First of all, as Stanford (2006) points out, it doubtful that these two versions of Newtonian mechanics are really distinct theories, as required by UTE, rather than the same theory with a different, added conjunct. Second, since in either version of Newtonian mechanics the vast majority of propositions are identical, realist commitment to one version would thus not be substantially challenged if the other version turned out to be correct instead. Third, realists should commit to the truth of only those theoretical claims 'that our theories themselves imply are amenable to empirical investigation' (16). Newtonian mechanics per se does not imply anything about the speed of the universe, and thus 'realism should no more extend to the conjunction of Newtonian theory with claims about the absolute velocity of the universe than with claims about the existence of God' (16).

There are other alleged examples for UTE, such as special relativity versus a classical mechanics in which forces act on measuring rods and clocks so as to simulate time-dilation and length-contraction and Bohmian hidden-variable versus standard von Neumann–Dirac formulations of quantum mechanics. These cases are controversial. But even if any of these examples could be convincingly argued to support UTE, it is not clear they could do much to undermine realism. After all, UTE is a *general* thesis that is supposed to be a threat for *any* realist commitment, not just commitments to some particular theories, mostly from physics (Laudan and Leplin 1991; Stanford 2001, 2006).

Stanford (2006), in his much noticed book *Exceeding Our Grasp*, has given new impetus to the debate about UTE and the realism debate more generally. In that work, Stanford articulates what he calls the *problem of unconceived alternatives* (PUA):

> I am suggesting, that is, that any real threat from the problem of underdetermination comes not from the sorts of philosophically inspired theoretical alternatives that we can construct parasitically so as to mimic the predictive and explanatory achievements of our own theories, but instead from ordinary theoretical alternatives of the garden variety scientific sort that we have nonetheless simply not yet managed to conceive of in the first place. (17–8)

Stanford's idea is as follows: in the history of science, we have repeatedly failed to conceive of radically different theoretical alternatives to our former best theories that eventually turned out to be just as well (and even better) confirmed as our current best theories.

> For example, in the historical progression from Aristotelian to Cartesian to Newtonian to contemporary mechanical theories, the evidence available at

the time each earlier theory was accepted offered equally strong support to each of the (then-unimagined) later alternatives. (19)

The implication is of course that realists have no better reasons to believe in the unobservable entities postulated by our current best theories than they have reasons to believe in the (radically different) unobservables postulated by yet unconceived theories. They should thus withhold committing to the truth of their current best theories.

PUA is a transient form of UTE. That is, in contrast to traditional forms of UTE, it allows for the possibility that, over time, the evidence will tell for one theory and against others.[32] Yet PUA is still a threat to the realist because it has occurred *repeatedly* in the history of science, in all realms of science (not just physics). In contrast to UTE, the threat from PUA is concrete: it is generated from what has *actually* happened in the past. And in contrast to the rather limited support base of the UTE, PUA's support is sizable: it derives from all those realms of science where empirically successful theories were at some point replaced by theories which accommodated the same evidence (and more) and which postulated radically different unobservables. PUA is therefore closely related to so-called pessimistic meta-induction, which we will consider in Chapter 2, where we will also further discuss PUA. Let us just briefly mention here one specific worry about PUA: One may contest whether Stanford's PUA actually improves on the UTE in terms of being more than just a Cartesian fantasy. Magnus (2010), for example, claims that special relativity was as remote a possibility in 1785 as that we are all just brains in a vat, because at that time no phenomena were known that would have suggested relativistic effects. We therefore do not have to take PUA more seriously than Cartesian fantasies envisaged by the defenders of UTE. However, the crucial difference between PUA and Cartesian fantasies, as Egg (2016) points out, is that we do have evidence supporting PUA in the form of past failures to conceive of better theories; we do not have such evidence with regard to our belief that the world we live in is not real (and generated by a computer connected to our brains). Again, there is much more to be said about PUA in Chapter 2, where we will seek to clarify its relation to the pessimistic meta-induction, but this discussion must suffice for now.

---

[32] Stanford thus accepts a point against UTE made by Laudan and Leplin (1991), who argue that our theories are always tested together with context-specific auxiliaries which might not be suited to future contexts. So, what may be an example of UTE at one point in time might no longer be such at a later point.

1.3   Theoretical Virtues: Epistemic or Pragmatic?

As we saw previously, both the local argument and the no-miracles argument for realism appeal to IBE, which in turn presupposes that it can be determined what the best explanation is. That requires criteria, at least some of which coincide with the criteria for good scientific theories, namely theoretical virtues, which we considered in the first part of this chapter. A central bone of contention between realists and antirealists is whether theoretical virtues are *epistemic* or merely pragmatic, i.e., whether these criteria are indicative of a theory's truth, also with regard to unobservables, or whether they are merely virtues with regard to the (convenient) *use* of a theory. For example, is simplicity a sign of truth or merely a feature that facilitates convenience? This question has several ramifications. First, if theoretical virtues were epistemic, belief in a theory on the basis of these virtues would be justifiable and thus rational. In contrast, if they were merely pragmatic, we may well pursue theories on such a basis – i.e., we may well explore and develop a theory and what follows from it – but we wouldn't be justified in believing in them (see also Chapter 6). Second, if theoretical virtues were epistemic, the use of inferences appealing to those virtues would be justified. This concerns both IBE and NMA, which, as we saw, employs IBE. Third, if theoretical virtues were epistemic, the threat posed by UTE could be fended off: the empirical equivalence of two theories could be broken in cases in which one of the theories considered was more virtuous than the other.[33] In what follows, I shall argue that perhaps the most controversial theoretical virtue of simplicity is indeed an epistemic one.

Some theoretical virtues are accepted as epistemic by both realists and antirealists. Most obviously, an antirealist like van Fraassen accepts that empirical accuracy (accommodation of the facts at *this* point in time) is indicative of a theory's empirical adequacy, i.e., indicative of a theory's truth concerning the observables in present, past, and future. Other theoretical virtues that van Fraassen accepts as epistemic are external and internal consistency and empirical strength. A theory that is not externally consistent – that is, consistent with other empirically accurate theories – will thereby violate the theoretical virtue the antirealist holds the highest, empirical accuracy. And a theory that is not internally consistent will be vacuously true with regard to any observations (since anything follows

---

[33] The question of whether and, in particular, which theoretical virtues are epistemic is also currently discussed in the *inductive risk debate*, where it is deliberated whether or not non-epistemic virtues may be allowed to influence our epistemic attitudes. See, e.g., Douglas (2009); cf. fn. 35.

from a contradiction), which is not acceptable to an antirealist, for she plausibly also embraces the idea that a good scientific theory must be testable. Van Fraassen says rather little about what he means by empirical strength, but he appears to think of it in terms of informational content, whereby an empirically stronger theory is one which is inconsistent with more facts than is an empirically weaker theory.[34] Antirealists have also had curiously little to say about fertility, and in particular novel success. It seems, though, that for them fertility represents no special virtue over and above the empirical adequacy of a theory. Likewise, antirealists – contrary to realists – tend not to emphasize that theories worthy of belief ought not to be ad hoc.

As we have seen earlier, when it comes to the scope of a theory, we should distinguish between empirical scope and theoretical unifying power. The former can be subsumed under empirical adequacy. Unifying power, on the other hand, and simplicity have been explicitly singled out by van Fraassen as non-epistemic virtues (van Fraassen 1980, 87–90). A very common criticism of viewing simplicity as an epistemic virtue concerns its *ontological* justification: we have no (and cannot have) any evidence for reality being simple. It is therefore mistaken to believe that simple theories are more likely to be true than complex ones (cf. van Fraassen 1980, 90). Simplicity and unifying power, for van Fraassen, are just pragmatic virtues, i.e., they are virtues when it comes to the use of theories, not when it comes to belief. A theory that is articulated in simpler mathematics, for example, is easier to handle than one that comes with a more complex one. Simplicity concerns, van Fraassen submits, are 'specifically human concerns, a function of our interests and pleasures' (87). Yet he is adamant that 'pragmatic virtues do not give us any reason over and above the evidence of empirical data, for thinking that a theory is true' (4). Accordingly, these virtues 'cannot rationally guide our epistemic attitudes and decisions' (87).[35]

Lycan (2002) has criticized attacks on simplicity that presume an ontological justification as misconstruing the target: Theoretical virtues

---

[34] More precisely, van Fraassen defines empirical strength semantically in terms of classes of empirical substructures of theories: the empirically stronger a theory, the fewer empirical substructures (or models) it possesses (van Fraassen 1980, 67–8). That is, the empirically stronger the theory, the more facts there are that are inconsistent with it. Van Fraassen's empirical strength thus seems to map onto Popper's idea of degrees of testability.

[35] Similarly, Douglas (2009) has argued that although theoretical virtues and social values should influence scientific decision-making (e.g., in statistical hypothesis testing), when the research in question has social impact, only epistemic virtues such as consistency and empirical accuracy ought to influence our epistemic attitudes towards our theories.

set *epistemological* norms for preferring theories. But such norms are independent of the ontology of the world. Consider that the norm to use a sharp chisel when sculpting marble does not at all require that the marble be 'sharp' in any way. Still, a sharp chisel will do us much better service in carving marble than a blunt one. Likewise, simplicity might well help us identify correct scientific theories, even when reality itself is complex. For example, Newtonian mechanics is simple in that it is built on three fundamental laws (plus the law of universal gravitation). But the reality of gravitational interactions between material bodies, as in our solar system, can be quite complex. The gravitational attraction of the moon by the sun and the earth, for example, is a three-body problem and has no analytical solution. Also, many motions in this world are complex in that they are composed of more elementary ones (such as straight-line motion and gravitation combining for parabolic motion). The general point pertains not only to syntactic parsimony, but also to ontological parsimony. Say you have a theory that postulates 32 different types of particles and another one that postulates 61. Could it be correct to say that a preference for 32 particles is justified by the world being simple? As we saw previously (Section 1.1.5), however, there is not likely to be any meaningful *absolute* preference for simpler theories. Rather, we need to assess our theories on the basis of simplicity by testing whether the entities postulated by a theory are needed to explain the data in question. So it might well be that we cannot do with less than 61 entities. The world justifies our preference for simplicity only insofar as the theory postulating fewer entities is better confirmed by the evidence than one which postulates more entities. Overall, the criticism regarding the ontological justification of epistemicity of theoretical virtues thus seems to be simply wrong-headed. Still, the realist owes the antirealist an argument as to why theoretical virtues, and in particular simplicity, ought to be epistemic. There are only a handful of proposals, which we shall now consider.

One obvious way to try to establish that simple theories are likely to be true would be to seek to establish that, as a matter of empirical fact, it has been the case (so far) that certain theoretical virtues such as simplicity have regularly, or even more often than not, resulted in (approximately) true theories throughout the history of science (cf. Salmon 1990 and Psillos 1999, 172). Yet such a project is beset with difficulties. First, what we could show empirically is that simple theories have regularly led to empirical success. But to argue that simple theories have regularly resulted in *true* theories (rather than just empirically successful ones) would require a further step – a step in fact that cannot be made empirically. As we have seen earlier in our discussion, usually this step is made by means of the NMA, whose soundness the antirealist denies. Even the first step, however, has not successfully been taken hitherto. In fact, it has not even been attempted. And where would one start? Which theories would one include in this induction? Could one exclude any theories at all? A fully comprehensive induction seems practically impossible: it would have to include any theory ever proposed. It would seem that the justification of simplicity as an epistemic virtue would require a more principled argument.

Forster and Sober's (1994) rationale for the epistemicity of simplicity, which we briefly mentioned earlier, seems more principled than the empirical approach we just considered. However, as we pointed out, the soundness of their rationale appears to be restricted to data models; there is no reason to believe that simplicity is positively correlated with predictive accuracy in higher-order theories. Furthermore, their account does not really provide what the realist needs, namely an argument for simple theories being more likely to be true (rather than just predictively more accurate).

Swinburne (1997, 2001) has suggested that 'it is an [synthetic] a priori epistemic principle that simplicity is evidence for truth' (1997, 1). Swinburne believes that we are forced to believe so because, despite the fact that there seems no *a posteriori* justification for simplicity being epistemic (it is highly doubtful that we can establish that simple theories turn out to be more empirically successful than complex theories, as mentioned before), and despite the fact that it is no analytic truth that simplicity is a successful guide to truth or probable truth (e.g., it does not follow from the probability calculus), we rely on it *successfully* in our epistemic practices in our everyday lives and in science (e.g., in curve-fitting) (Swinburne 1997, 44ff.). Swinburne concludes that '... either science is irrational (in the way it judges theories and predictions probable) or the principle of simplicity is a fundamental synthetic a priori truth' (56).

Sober (1990), in contrast, has argued that simplicity 'does not have an *a priori* and subject matter neutral justification' (152). According to him, any apparent 'global' appeal to some general criterion of simplicity *reduces*, in reality, to merely 'local' – i.e., domain-specific – substantive considerations. In Sober's discussed example of phylogeny, one way of constructing phylogenetic trees proceeds by choosing trees with the fewest evolutionary character state changes. Although one may think that phylogeneticists are appealing to some general simplicity constraint, in reality, their procedure is justified by the assumption that evolutionary change is unlikely. Since, for Sober, simplicity considerations are local, they will differ from one context to another. In fact, Sober believes that 'What makes parsimony reasonable in one context therefore may have nothing in common with why it matters in another' (Sober 1990, 140). Thus, the self-proclaimed reductionist Sober is also a contextualist about simplicity.[36]

Sober's account of simplicity presents a puzzle: if it is true that simplicity concerns reduce to concerns about substantive and domain-specific assumptions, and if there is no global simplicity criterion, what role does simplicity really play? Does it play any role at all? If not, then Sober's account looks like an extreme form of reductionism, namely eliminativism, according to which we might as well abandon any talk of simplicity. On the other hand, if the local concerns really are simplicity concerns, then there must be something about any of those local concerns that makes them recognizable as simplicity concerns. Thus, on this view, there would have to be a 'global' notion of simplicity after all. So either Sober's account eliminates simplicity concerns or it *does* require a global notion of simplicity. In the face of the widely reported use of simplicity in theory choice in science,[37] the former option is unattractive.

1.4 The First Virtuous Argument for Realism: The Argument from Simplicity

As I pointed out previously, a good account of simplicity considerations in science *is* one that allows for the context sensitivity of such considerations. That is, in some contexts, one form of simplicity might be deemed as most relevant by the scientific community when it comes to choosing a theory, and in other contexts, others. The guiding principle for all these forms of simplicity considerations is plausibly what I have called the evidential-explanatory rationale (cf. Section 1.1.5): an entity or principle postulated by a theory can receive empirical support only insofar as it is required to explain the evidence. If a theory could explain the same set of data without the entity or principle in question, that entity or principle should count as unsupported by the data. Accordingly, one ought to postulate only as many entities or principles as are supported by the evidence when explaining the relevant set of data. Simplicity considerations thus come down to

---

[36] Cf. Sober (2015). Another reason Sober refers to himself as a reductionist is that 'parsimony is never an epistemic end in itself' (199). This statement receives meaning also from Sober's joint work with Malcolm Forster. See earlier references to this work and Chapter 3.
[37] See McAllister (1999), Baker (2013), and Fitzpatrick (2013).

explanatory-evidential considerations. What does that mean for the realism debate?

Take again two theories T_1 and T_2 and suppose that they both save the phenomena. Suppose further that T_1 is simpler in that it postulates fewer entities or principles than T_2 (for simplicity's sake, I'll focus on entities in what follows, but my point applies equally to principles). Since for the antirealist simplicity is merely a pragmatic virtue, this difference is epistemically insubstantial. The evidential-explanatory rationale, however, tells us that the difference is epistemically significant indeed: since T_1 accomplishes the same explanatory task as T_2, the extra entities postulated by T_2 are empirically unsupported. The evidential-explanatory rationale thus compels us to choose T_1 over T_2, *despite* their empirical equivalence. By virtue of simplicity considerations, we have better reasons to believe in T_1 than in T_2. But if a theory's degree of confirmation depends in this way on whether or not a theory is simpler with regard to the theoretical entities it postulates, then simplicity is clearly an epistemic concern. This is bad news for the antirealist, who has it that a theory's saving of the phenomena is our *only* reason for belief in the theory. This is my first argument for realism, namely the *argument from simplicity*.

As we have seen in the preceding pages, according to the evidential-explanatory rationale, simplicity is a context-dependent theory-choice criterion: it does not apply in contexts where there are theories that postulate fewer entities than needed to explain the phenomena. In such contexts, more-complex theories may indeed be preferable. They would be, that is, if more-complex theories postulated entities that are all needed to explain the phenomena. The realist's commitment to the epistemicity of simplicity should therefore not be blind.

We also noted that the *form* of simplicity must be specified before theories can be compared. Luckily, there appear to be only two broad forms of simplicity in actual (high level) theory choice, namely syntactic and ontological parsimony. Which of those two forms will be relevant depends on the particular context in which the theory in question is adopted by scientists. Philosophers who choose to be (local) realists about these theories can leave the choice of the form of simplicity to the scientific community, as they are best suited for deciding which form of theoretical simplicity should matter in any particular context of *scientific* theory choice. With regard to the more global realism debate, in which we abstract from any particular theory, a similar strategy is in order. After all, realists want to commit themselves to all those theories which the scientific community determines to be the best we have of reality. Likewise, realists should commit themselves to those *forms* of simplicity which scientists select in order to determine their best theories. However, the best *rationale* for simplicity considerations, namely the evidential-explanatory rationale, is of course still independent of what the scientific community might think about it.

There is a further point that should be clarified. It concerns our epistemic commitment to the entities determined as (explanatorily) super-fluous by the application of the criterion of simplicity: we can either *deny* their existence, or we can *suspend* our belief with regard to their existence, i.e., we neither believe in their existence nor disbelieve it.[38] Clearly, the denial of the existence of explanatorily superfluous entities would be foolish: the fact that our theory can do without a particular entity when explaining a certain set of phenomena does not rule out that this entity may be required for the theory's explanation of *other* phenomena or that it may be required by *other* theories successfully explaining other phenomena. The denial of the existence of explanatorily superfluous entities is thus straightforward fallacious reasoning.[39]

As mentioned earlier, the other theoretical virtue van Fraassen claims to be merely pragmatic is unification. As we have seen, unification is closely related to simplicity: the higher the unifying power of a theory, by the lights of leading accounts of unification, the simpler the theory should be with regard to the number of argument patterns, basic phenomena, fundamental assumptions, etc., which the theory stipulates. So insofar as simplicity is an epistemic virtue, unification is so too. Are there indepen-dent grounds for unification being an epistemic virtue? Obviously, the more facts a theory accommodates, the more likely it is that the theory is true. But as we mentioned previously, we need to distinguish between empirical scope and theoretical unification. The former virtue is consistent with an antirealist stance. Is theoretical unification an epistemic virtue? It is striking that many of the theories the realist believes to be approximately true do unify the phenomena theoretically: for example, the standard model of particle physics (unifying the weak, the strong, and the electro-magnetic force), Newtonian mechanics (unifying terrestrial and celestial

---

[38] This distinction was originally drawn by Sober (1981) and baptized 'atheistic razor' and 'agnostic razor', respectively. Sober (2015) uses the terms Ockham's razor of denial versus Ockham's razor of silence for those two ideas. The distinction is discussed in depth by Barnes (2000).

[39] Sober (2015) argues that Ockham himself subscribed to the stronger reading. On that basis, Sober in fact rejects what I have called the evidential-explanatory defence of simplicity. Contra Sober, I do not think we must follow what Ockham himself thought about his principle when pondering whether the application of Ockham's razor (in some form) is reasonable.

mechanics), quantum electrodynamics (unifying the interactions of light and matter), quantum mechanics (unifying wave and particle character-istics of light and matter), the theory of general relativity (unifying inertial with gravitational phenomena). The amount of new facts these theories engender is often outweighed by the theoretical advance they offer. For example, the three classical tests of the general theory of relativity are light-bending, the advance of the perihelion of Mercury, and gravitational redshift. Until the late 1950s, no robust data were available on the latter two predictions (Will 1993). And the early confirmation through light-bending turned out to be much less unequivocal than normally portrayed (see Section 6.2.1). The realist can of course argue that the empirical success, and in particular the novel success, of these theories is miraculous unless they get something fundamentally right about nature. Arguing from empirical success to truth is what the realist has traditionally done. But it might also be argued that a theory is likely to be true when it theoretically unifies. At the very least, that would explain physicists' early espousal of the general theory of relativity despite its lagging (and numerically limited) empirical success.

CHAPTER 2

Pessimism, Base Rates and the No-Virtue-Coincidence Argument

In Chapter 1 we saw that realists have appealed to the explanatory virtues of scientific theories in order to fend off the antirealist threat from the underdetermination of theories by the evidence. Regardless of what one might think about this defence, there is another antirealist challenge that has received much more attention in recent years: the so-called pessimistic meta-induction (PMI). In Sections 2.1 to 2.3 of this chapter, we will review and assess the current status of the debate surrounding the PMI and its cousin, the problem of unconceived alternatives (PUA). I believe that neither the PMI nor the PUA, in their current form, pose any substantial threat to realism.

Both realists and antirealists have recently been accused of neglecting the base rate of true theories (Section 2.4). The dependence of the central arguments of realism and antirealism on the base rate of true theories, which are inaccessible to us, makes this a particularly severe challenge. I will confront this challenge in Section 2.5 by developing a new type of convergence argument, namely the *no-virtue-coincidence argument*, which forms my second, and central, virtuous argument for realism. I will conclude the chapter in Section 2.6.

2.1 The Pessimistic Meta-Induction and the Divide et Impera Move

The PMI, according to Laudan (1981), is an inductive argument against realism on the basis of a list of past theories which had some empirical success but whose ontology nevertheless turned out to be utterly mistaken. Laudan's list – which he famously claimed could be extended ad nauseam – includes the following items: Aristotle's theory of crystalline spheres of the heavens, Galen's humoral theory of medicine, the effluvial theory of static electricity, the phlogiston theory, the caloric theory of heat, the vibratory theory of heat, the vital-force theory of physiology, theories of spontaneous generation, and ether theories. These theories were clearly utterly different from our current ones. None of our current respected theories postulates anything resembling a substance for heat, a 'principle' that makes materials combustible, and we do not believe that we get sick when our four bodily fluids are in imbalance. Moreover, we know that these ideas are utterly mistaken, and we have replaced them with ones that have resulted in much greater empirical success.

As has been pointed out by several authors (e.g., Psillos 1999; Saatsi 2005a), the PMI is a *reductio* targeting the no-miracles argument (NMA), introduced in Chapter 1 (Section 1.2.3). The PMI can be reconstructed in this way:

P1: The only explanation for our current theories being as empirically successful as they are is that they are (approximately) true.

P2: There were many theories in the past, considerably different from our current ones, which were empirically successful too.

P3: But these theories were clearly false; their central theoretical terms did not refer to anything real in the world.[1]

C: Hence, in contradiction with P1, truth cannot be the right explanation for a theory's empirical success – neither for our past nor our current or future theories' success.

Let us first consider some clarifications. Sometimes the PMI is read differently. Instead of inferring C, some read premises P1–P3 to imply that our current (and future) theories are *likely* to be false (e.g., Lange 2002; Magnus 2010; Fahrbach 2011). At the very least, however, such an interpretation is at odds with antirealism as defended by its main modern proponent, van Fraassen (cf. Chapter 1): The modern antirealist is a sceptic, not a dogmatist. He does *not* claim that our theories are *in fact* or *likely* to be false, but rather that they *might* be false, for all we know. So either this interpretation of the PMI is simply mistaken[2] or the antirealism defended with this interpretation of the PMI is a different (much less plausible) type of antirealism.[3]

---

[1] Reference here is to be understood in the sense of the correspondence theory of truth (cf. Section 1.2.1). Some philosophers have sought to develop responses to the PMI by the means provided by the philosophy of language, and in particular, by way of the causal theory of reference. This debate has its own quirks and twists but will not concern us here. For more details, see for example Psillos (1999) and Stanford (2006).

[2] See also (Saatsi 2005a) for a similar critique of the foregoing alternative interpretation of the PMI, and in particular of Lange (2002).

[3] Antirealists of course accept premise P1 only for the sake of the argument. As we saw in Chapter 1, they have offered alternative explanations (with debatable success) or denied that the success of science needs to be explained in the first place.

The PMI presupposes two further (implicit) assumptions: first, that there must be one and the same (i.e., universal) explanation as to why our current and past theories are successful and, second, that the success of our current and our past theories is relevantly similar so as to require the same explanation. The first assumption is typically, but not always, accepted by realists. That is, normally it is accepted that different, but similarly successful, theories should be explained in the same way, rather than in a more piecemeal fashion. We will come back to this in Section 2.3. Realists are however much less willing to grant the second implicit assumption, namely that the success of our past and our current theories is relevantly similar. The empirical success of our past theories, they argue, is indeed very different from the success which our current theories continue to impress us with. More specifically, realists have pointed out that many items on Laudan's list never had any *novel* success; i.e., they never predicted any novel phenomena, but merely explained extant ones. But for us to make a realist commitment to a theory, the argument continues, a theory must successfully predict novel phenomena. This requirement is usually motivated by saying that explanatory success is 'too easy' (Psillos 1999, 104ff.). We shall put this idea under detailed scrutiny in Chapter 3. Let us simply assume here, with the vast majority of commentators, that once the novel success criterion is considered, Laudan's list reduces to only a couple of items (most notably, ether theories and the caloric theory of heat). But, as mentioned earlier, Laudan believed that his list of successful but false theories could be extended at will. And indeed many false theories have since been identified that have managed to produce *novel* success. For example, Dirac successfully predicted the existence of the positron based on the entirely false idea of the vacuum as an infinite 'sea' of particles, in which positrons would figure as 'holes' (Pashby 2012). Other examples of false-theory novel success pairs include these: the spurious Titus–Bode law (together with the laws of Newtonian mechanics) and the planet Neptune (McIntyre 2001; Lyons 2006), Ptolemy's geocentric system and various regularities in planetary orbits (Carman and Díez 2015), the caloric theory of heat and the speed of sound (Psillos 1999; Lyons 2006), the phlogiston theory and redox reactions (Carrier 1991; Lyons 2006; Ladyman 2011; Schurz 2011), the Bohr–Sommerfeld model and the fine structure of the hydrogen atom (Vickers 2012), and the wave theory of light and Kirchhoff's diffraction formula for light passed through an aperture in a screen (Saatsi and Vickers 2011).[4] Realists have gone to great lengths to

---

[4] These and many other cases (altogether 20) are conveniently summarized by Vickers (2013a).

accommodate such cases. Most importantly, they have become more nuanced in their commitment to the truth of novel-success-generating theories. Rather than inferring the truth of the *entire* theory which produces novel success, they have distinguished between those parts of a theory that are essential in the derivation of a theory's novel success and those that are not. It is only the former parts that realists have typically deemed worthy of belief.[5] There are several ways of dividing up theories: into 'working posits' and 'presuppositional posits' (Kitcher 1993), 'structure' and 'content' (Worrall 1989c), 'idle constituents' and 'essentially contributing constituents' (Psillos 1999), 'detection properties' and 'auxiliary properties' (Chakravartty 2007), 'derivation-internal posits' and 'derivation-external posits' (Vickers 2013a), etc.[6] Once theories are so divided, ideally a continuity can be established between present and past empirically successful theories *despite* the fact that our present and past theories conflict substantially. Perhaps the most famous example for which a continuity of novel-success-fuelling theoretical parts has been argued is the shift from Fresnel's wave theory of light to Maxwell's electromagnetism (Worrall 1989c). Although Fresnel formulated his theory on the basis of the assumption that light would propagate in wave-like motions through a 'luminiferous' ether, this assumption was later given up. Intriguingly, Fresnel's equations, however, were retained in this theory change. Worrall uses the case to argue for structural realism, viz. the view that we should believe only in the structures our theories posit to explain the phenomena, and more specifically those structures which are conserved under theory change and which are responsible for the theory's novel success.[7]

The divide et impera move, as the realist's dividing up of theories is also known, has been criticized on various fronts. Some have argued that particular ways of dividing up particular theories, such as Fresnel's ether theory along a structure/content divide, is not warranted and that the parts that allegedly fuelled the theory's novel success on their own did not do so

---

[5] Saatsi (in press) sets straight those who have claimed that a theory's *explanatory* success is sufficient for realist commitment.

[6] Harker (2013) proposes that realists, when applying their *divide et impera* move, turn their attention away from (absolute) empirical success to *progress*, understood as comparative empirical success: a theory is successful when it accounts for phenomena which rivals cannot. See Chapter 3 for more details about comparative success. See also fn. 8 in the current chapter for more on Harker's notion of progress and realism.

[7] Ladyman (1998) pointed out that Worrall's formulation is ambiguous between an epistemic and an ontic version of structural realism. Whereas the former has it that we should believe only in the structure of theories, the latter says that the structure is all there is to reality. Ladyman and others have adopted the latter, stronger form of structural realism, while Worrall has since then sided with the epistemic form.

(Psillos 1999; Saatsi 2005b). Others have insisted that despite some limited success of the divide et impera strategy, there are still numerous cases that the realist hasn't accommodated yet (Lyons 2006; Vickers 2013a). The most influential and far-reaching critique, however, is the one advanced by Stanford (2006). In order to determine which parts of our past theories are (i) truth-candidates and (ii) success-fuelling, he points out, realists cannot but make reference to our *current* theories. Only with the benefit of hindsight provided by our current theories can we decide which theory parts should now be considered false and which ones can be considered to have contributed to the success of current theories. The upshot of Stanford's critique, then, is that the realist's divide et impera procedure is inherently biased towards what realists want to establish. Stanford concludes that

> it is the very fact that some features of a past theory survive in our present account of nature that leads the realist *both* to regard them as true *and* to believe that they were the sources of the rejected theory's success or effectiveness. So the apparent convergence of truth and the sources of success in past theories is easily explained by the simple fact that both kinds of retrospective judgements have a common source in our present beliefs about nature. (Stanford 2006, 166)

To illustrate his point, Stanford invites us to apply the divide et impera to cases where a false theory F1 is followed by another false but nevertheless success-generating theory F2, which we believe to be true on the basis of (a) F2 producing success and (b) F2 succeeding F1. Would we not be inclined to judge only those parts of F1 to be truth-candidates that we also find in F2? And would we, likewise, not also be biased towards deeming success-fuelling only those parts of F1 that we find in F2? Stanford points out that even theoretical posits shared by F1 and F2 and fuelling the success of the respective theories may in fact be wrong. For example, both Darwin's theory of pangenesis and Weismann's (successor) theory of germ plasm shared the false posit that hereditary particles would be specialized for the expression of only certain kinds of cells in subsequent generations. Stanford concludes that the continuity between subsequent theories in the history of science which the realist identifies via the divide et impera is 'virtually guaranteed', independently of whether our current theories are actually (partially) true. In other words, finding continuity is too cheap and can therefore not be the basis for an argument for realism. Stanford believes that a successful defence of realism would require the realist to provide *prospective* criteria for selective realism, so that the divide et impera move could be used to make predictions about which theory parts will be retained in the future development of science.

In response to Stanford's challenge, realists have pointed out that realist commitments usually require theories to be capable of generating not just any old success, but *novel* success: it looks as though many of the past theories that Stanford discusses in his book did not enjoy this capacity. More importantly, the question of whether or not a theory has had (novel) success is an empirical one and can be given a correct answer independently of any consideration of our current theories (Psillos 2009; Votsis 2011; Peters 2014; Saatsi, in press). Still, although Stanford is clearly mistaken when he, for instance, claims that retrospection 'virtually guarantees' that one will find what one is looking for (166), the possibility of bias towards selecting favourable theory bits and disregarding others must be taken seriously. For example, whilst the realist might entirely correctly identify elements in past theories that were responsible for their novel success and link those to current theories, she may, because of her bias for establishing harmony between past and current theories, simply overlook parts of theories that were responsible for producing novel success without being carried over to current theories. Perhaps there is a more independent procedure? Psillos (1999, 110) suggests that a theory part has a legitimate claim to fuelling the novel success of theories if and only if it cannot be replaced by an alternative posit which (i) is consistent with the general theory, (ii) would preserve the empirical content of the theory, and (iii) would avoid making the theory ad hoc. Yet, as Lyons (2006) points out, these criteria are neither relevant nor fit for use. They are irrelevant because there could exist an alternative posit for the actual posit that satisfies (i)–(iii), and yet that would not have any bearing on whether or not the actual posit figures in the derivation of the novel success in question.

In addition to the conditions (i)–(iii), Psillos proposes that we consult scientists themselves when applying the divide et impera move and ask whether they themselves believed the contentious entities, such as the ether, to be true. When they do not, this can give us good indication that the entities in question did not essentially contribute in the production of novel success (Psillos 1999). In particular, Psillos cites various sceptical comments about central theoretical posits by ether theorists, including Maxwell, and by theorists postulating caloric. But Psillos's proposal is problematic: not only are scientists perhaps not the best individuals to consult about this question (simply because they generally do not care very much about the realism debate), but also one is bound to find conflicting judgements, if one finds any. Indeed, as Stanford points out, scientists have repeatedly been utterly mistaken in their beliefs about theoretical posits. In particular, and somewhat ironically, Maxwell was quite robust in his belief in the ether, i.e., an entity which Psillos claims can be shown to be dispensable by attending to what scientists say (Stanford 2006, 152ff.).

In order for the realist not to be biased by our current theories towards picking out only congenial parts of past theories, Stanford demands that realists provide an explicit criterion that can be applied prospectively to future theories (168). But must the realist really provide such a criterion? With Saatsi (in press) we may ask whether such a demand would really be reasonable: it would ask the realist to provide criteria that *scientists* could use to identify parts of their theories that would be likely to survive theory change. The realist would, for example, be able to tell the Newtonians prior to 1915 which parts of their theories would be likely to survive after the Einsteinian revolution that would bring about the special and general theory of relativity. But not even scientists were able to do this. So why should the realist be held up to such high standards? Instead, Saatsi argues, any kind of 'recipe' for identifying novel-success-fuelling posits that is suitable for all cases should be given up. Not all realists dismiss Stanford's challenge. Votsis (2011, 1231), for example, accepts the challenge and argues that the structural realist *already* provides the criterion Stanford is calling for: the structural realist predicts that the theory that might supersede quantum electrodynamics will preserve 'at least in some limit' the equations of Fresnel, Maxwell, and quantum electrodynamics.[8] A problem that structural realism has faced (Psillos 1999, 159; Saatsi 2005b; Stanford 2006), however, is that the preservation of equations throughout theory change is simply not enough for realism: the equations need to be interpreted substantially, and not just minimally (as Votsis suggests), for realism not to collapse into anti-realism. But once one allows for a more substantial interpretation of the equations, the divide between structure and content is hard to maintain.[9]

---

[8] Harker (2013) claims that a realist divide et impera strategy that focuses on empirical progress rather than empirical success meets Stanford's demand for a prospective criterion. For Harker, those theory parts will be retained which are responsible for progress. But it is not clear how this constitutes an improvement over the standard realist's prediction that those theory parts will be retained which are responsible for their novel success. Neither account offers a principled way for identifying the relevant theory parts prospectively. But again, it is questionable whether the realist is obliged to provide such principles. See also fn. 6.

[9] A useful overview of structural realism and objections to it can be found in Frigg and Votsis (2011).

2.2   False Posits and Novel Success: What If?

Let's suppose that the antirealist was able to find theories that were empirically successful but false and that she could show beyond reasonable doubt that false assumptions did fuel those successes. What would follow? Interestingly, there is some confusion in the literature about this question. As Saatsi and Vickers (2011) have pointed out, several scientific realists believe that a single such theory, or just a handful of them, would undermine realism (e.g., Leplin 1984; Psillos 1999). Saatsi and Vickers describe this view as 'naïve optimism'. In contrast, they believe that a more appropriate form of realism should be able to tolerate such theories by allowing that there might be explanations other than truth for a theory's success. Such explanations, they surmise, could have to do with 'the domain of theorizing in question', 'the mathematics used in the derivation', 'the nature of the system under theorizing, or by anything of that ilk that might conceivably power success-production under some particular circumstances' (33).

Saatsi and Vickers present a theory of the kind just mentioned, namely Kirchhoff's derivation of the correct diffraction formula for light passing through an aperture in a screen. Kirchhoff, Saatsi and Vickers argue, made so 'wildly wrong assumptions' that the case cannot be accommodated by the realist with a divide et impera move (42).[10] Although they themselves find it hard to figure out how Kirchhoff's success can be explained, they invite the realist to try to 'isolate' this case by 'showing how the field of theorizing in question is idiosyncratic in relevant respects' (42). If this could be shown, then, Saatsi and Vickers surmise, the realist's standard explanation of novel success would not be in danger. But suppose the realist can't isolate Kirchhoff's case in this way and there is nothing idiosyncratic about it. Would that really be so bad for the realist?

Unless we are talking here about the naïve optimist type of realist, I don't think the realist has any obligation to identify a non-realist explanation of a false, novel-success-producing theory. In fact, I think

---

[10]   Vickers (2013a), revisiting his co-authored study of Kirchhoff's successful prediction, backtracks on the claim that this case is problematic for the realist on the basis of the *logical* point that any novel-success-producing posit 'might "contain within it" some other posit that is the real working part', whereby containment is to be understood as the possibility of inferring some logically weaker statement, such as 'Dick is too heavy' from 'Dick is exactly 10 pounds too heavy'. Even when the latter is false, the former can still be true (198). Likewise, a part of a theory that has been shown to be essential for deriving a novel prediction might itself turn out to be wrong and a logically weaker version of it nevertheless true. Vickers himself fails to demonstrate that this logical possibility materializes in the Kirchhoff case (206ff.). But unless a demonstration can be had, the burden of proof remains with the realist. The logical point won't suffice.

Saatsi and Vickers got the dialectic mixed up here: If there really was a non-realist explanation for why Kirchhoff got it right despite 'wildly wrong assumptions', then the realist would be in trouble. Because then it would no longer be a miracle that a theory managed to produce novel success, despite its falsity. Further, if there really is an antirealist explanation of the success of theories such as Kirchhoff's, the antirealist would have some legitimate hope of projecting her explanation to other cases of false theories that have generated novel success. In the absence of such a non-realist explanation, however, the non-naïve realist can rest content, admitting that there might be lucky flukes where false theories have (for mysterious reasons) managed to get things right, but that in the majority of cases, the success of theories is best explained by their approximate truth.[11]

Of course there are limits. If it should turn out that the vast majority of theories that have managed to produce novel success are false, and the *divide et impera* would fail, then the move discussed here would start to look untenable. The question must therefore be How can we possibly know how many such theories are out there? This proves a difficult question to answer. In the PMI, the antirealist can only list a *sample* of such theories. Likewise, the realist can only list a *sample* of predictively successful theories which have not turned out to be false, or theories to which the *divide et impera* move has been applied successfully. How can we know how typical such samples are? In other words, what does the underlying population of theories look like? Some philosophers have denied that we can give any reasonable answers to these questions. We will take up these issues again in Section 2.5. First, however, let us compare the PMI to the problem of unconceived alternatives (PUA), introduced in Chapter 1.

2.3 PMI versus PUA

Recall that the PUA has it that our current theories are underdetermined by theories which we haven't even yet managed to conceive. History tells us that we have repeatedly failed to conceive of theoretical alternatives that were as well supported by the evidence as the theories we held at a particular point in time. Obviously, the PUA and the PMI are similar. Both are inductions that appeal to the history of science, and more specifically to the fact that past scientific theories have turned out to be

---

[11] In a recent exchange with Held (2011), who believes that the mere possibility of a false theory producing novel success is enough to defeat the NMA, Morganti (2011, 2012) argues that the NMA is a probabilistic argument and therefore allows for such a possibility.

false with regard to the fundamental postulates they make about the world. But how do they differ, if at all?

Some have claimed that the PUA is just a 'novel red herring' that presents no substantially new challenge to realism than the PMI (Chakravartty 2008). If that is so, then the realist can simply employ the divide et impera move: although scientists in the past failed to conceive our current best theories, they did manage to conceive of theories that contained elements that we find in our current theories and that were responsible for their success (cf. Ruhmkorff 2011). According to Magnus (2010), however, the PUA is substantially different. First, whereas '[t]here is no way to treat all past theories as a population or pull a representative sample of them [as we would have to presume in the PMI] . . . it is at least coherent to treat past scientists as a population [as in the PUA]' (813). Magnus here appeals to his and Callender's statistical rendering of the PMI, which we shall discuss in more detail in Section 2.4. The PMI only has any force if it can be shown that the sample of successful but false theories figuring in the inductive base of the PMI is representative of the overall distribution of *all* successful theories (in the past and the present). For this we would have to know the base rate of true theories (see Section 2.4). In contrast, Magnus insists that the PUA is really an induction over scientists (and their 'favourite theories'), not theories (as in the PMI). And the set of all scientists is much better defined than the set of all theorists, or so Magnus argues (Magnus 2010). Yet, one may wonder whether this difference is really so substantial. After all, in the PMI, the theories in question are also conceived of by scientists, and in the PUA, it is still the historical track record of (at some point) unconceived successor *theories* that is supposed to undermine the NMA. But if that is the case, then the difference between the PMI and the PUA seems to boil down to difference in emphasis. Second, Magnus claims that although realists can try to at least weaken the threat posed by the PMI by arguing that past theories did not have the same kind of success as current theories (namely novel success), in the case of the PUA, realists would have to argue that past scientists were somehow different from current ones. So, Magnus continues, 'unless there is a reason to think that theorists have gotten better at exhausting the relevant possibilities', the PUA poses a stronger threat to realism than does the PMI (814). Indeed, on the basis of Kuhn's idea of normal science – according to which scientists are efficient in their work because they tend not to consider alternatives to the reigning paradigm – Magnus argues that scientists are in fact unlikely to exhaust the space of possibilities in their theorizing (815). In a similar vein, one may argue with

Stanford that the way in which modern research funding is structured makes it unlikely that non-mainstream theoretical alternatives are adequately explored (Stanford 2006, 132). Realists have countered that there have been significant improvements in the methods scientists are using in the past 50 to 100 years or so. The present is therefore relevantly dissimilar from the past, and the past record of science can therefore not be used to doubt the present (Roush 2010; Devitt 2011). The research community has also grown significantly bigger and has therefore had more resources to explore theoretical alternatives (Fahrbach 2011). In other words, scientists, by virtue of their improved methods, may indeed have become better at exploring theoretical alternatives. And peer review may, but need not, prioritize mainstream research. At any rate, instead of trying to settle this dispute here, let us focus on the more fundamental base-rate neglect charge.

2.4 Broader Concerns: The Base-Rate Fallacy

In an attempt to avert the threat of the PMI, some philosophers have surmised that false-yet-successful theories may just be a phenomenon of the past. Accordingly, we are not mistaken, contra PMI, to regard the novel success of our *current* theories as strongly indicative of truth. The PMI would then merely be 'evidence of the scarcity of true theories in the past' (Lewis 2001, 377). Although this is a logical possibility, whether or not that is the case would require us to properly assess the relevant reference classes – i.e., the class of false, true, successful, and unsuccessful theories – before we could evaluate whether there has indeed been a change in the distribution of those classes over time. None of this can be done easily. This difficulty also motivates part of a highly sceptical assessment of the entire realism debate by Magnus and Callender (2004): they accuse realists and antirealists alike of committing the so-called *base-rate fallacy*.

The base-rate fallacy can be illustrated with a simple example from the medical context. Suppose we were to test the presence of some disease T in a population of subjects with a very effective test. That test, let's suppose, would have a very high probability of indicating to us the presence of a disease when the disease is really present in a subject. Let us refer to a positive test result as *e*. Expressed formally, then, $P(e|T) \gg 0$. Suppose further that the test has a very low false positive rate. That is, the test is unlikely to indicate the presence of the disease when it is actually absent $(P(e|\neg T) \ll 1)$. For the sake of concreteness, assume that $P(e|T) = 1$ and $P(e|\neg T) = 0.05$. Contrary to intuition, it would then be fallacious to infer that the (posterior) probability of some subject having the disease, when the test indicates that the subject has the disease, is high, for example, $(P(T|e) = 0.95)$. In fact, it can be rather low. If the disease is very rare in the population (i.e., $P(T) \ll 1$) – for example, 1/1000 – given the presumed sensitivity of our test of $P(e|T) = 0.05$, we would expect 51 subjects in a population of 1000 to test positive. Since, by assumption, only one of those subjects actually has the disease, $P(T|e)$ would be just 0.02 – that is, much lower than the intuitive 0.95.

Magnus and Callender (2004), building on the observation by Howson (2000), accuse the partakers in the realism debate of having made the same mistake.[12] That is, they accuse realists and antirealists alike of having neglected the base rate of true theories in the pool of all theories, i.e., the prior probability of a (or any) theory being true. Instead, the debate has focused on the probability of a theory being false if empirically successful $P(\neg T|e)$ and the probability of a theory being successful if false, i.e., on the likelihood of e given $\neg T$ (i.e., $P(e|\neg T)$). As we've seen earlier, whereas antirealists have sought to increase $P(\neg T|e)$ with arguments like the PMI, realists have tried to decrease $P(e|\neg T)$ by restricting the notion of success to novel success (327).[13] Yet these probabilities are not very informative unless one also knows the base rate of true theories. It is not clear, though, how we possibly could. Expressed in Bayesian terms, the base-rate fallacy amounts to ignoring the dependence of the *posterior* probability of a successful theory being true on the *prior* probability of a theory being true. Accordingly, the realist and the antirealist can set their subjective priors as they please.

Although Magnus and Callender argue that their challenge is equally fatal to realists and antirealists, they pose the following dilemma to the realist only:

> Either there is a way of knowing the approximate base rate of truth among our current theories or there is not. If there is, then we must have some independent grounds for thinking that a theory is very likely true; yet if we had such grounds, the no-miracles argument would be superfluous. If there is not, then the no-miracles argument requires an assumption that some significant proportion of our current theories are [sic] true; yet that would beg the question against the anti-realist. (Magnus and Callender 2004, 328)

Magnus and Callender see no way out of this dilemma. What they call the 'wholesale' realism debate is simply irrational:

---

[12] Howson uses the base-rate fallacy only against the realist.

[13] $P(\neg T|e)$ and $P(e|\neg T)$ are related by Bayes's theorem:

$$P(\neg T|e) = \frac{P(e|\neg T)\, P(\neg T)}{P(e)}.$$

Without independent methods for estimating crucial base rates, there is little to do but make arguments that beg the question. Wholesale realism debates persist not due to mere stubbornness, but because there is *no reason* for opponents to disagree. (336; original emphasis) [14]

Although Magnus and Callender see no point in philosophers any longer engaging with the 'wholesale' realism debate – i.e., one about 'all or most of the entities posited in our best scientific theories' – they think that a 'retail' realism debate about 'specific kinds of things, such as neutrinos', can still be had on rational grounds (321).

Magnus and Callender's call for philosophers to attend to the potential truth of particular theories has been heeded by several philosophers (Psillos 2009; Saatsi 2009). However, there is a danger that 'retail' realists will end up finding it difficult to explain what it is that they do *in addition to* what scientists do. If retail realists refrain from appealing to wholesale arguments such as the NMA in arguing for the reality of atoms, neutrinos, etc., and just refer to the 'same evidence scientists use to support' their hypotheses about unobservables, then it would seem that there is no job left for the philosopher to do except watch scientists go about their business. But then the genuinely philosophical question about whether we have good reasons to believe that (some) of our best scientific theories are approximately true is simply left unresolved (cf. Dicken 2013).

Alternatively, philosophers may monitor scientists in their inferential practices and assess whether the available evidence *really* justifies the inference that, for example, neutrinos exist, drawing perhaps on philosophical theories of confirmation (as for instance suggested by Hanson 1962; cf. Chapter 7). Such a view would however grant philosophers more competence than scientists in assessing the truth of scientific theories, which seems presumptuous.

Despite the particularist approach being rather unappealing to the philosopher who wishes to contribute to the realism debate and to go beyond what scientists can – and are in a better position to – tell us about their theoretical posits, head-on confrontations with Magnus and Callender's challenge have been few and far and between.[15] I shall dare such a confrontation in the remainder of this chapter.

---

[14] Similarly, when the probabilities are interpreted as subjective probabilities, Magnus and Callender also 'can't imagine how one could find a reasonable set of priors' (329).

[15] See, for example, Psillos (2009) and a reply by Howson (2013). For couple more recent attempts, see Menke (2014) and Henderson (2015).

2.5   The Central Virtuous Argument for Realism: Virtue Convergence

My argument against Magnus and Callender will appeal to the Kuhnian picture of theory choice and proceeds roughly as follows: it would be an unlikely coincidence for scientists' judgement concerning the truth of a theory on the basis of the theory's virtues to converge if the theory wasn't in fact true. I refer to this argument as the *no-virtue-coincidence argument*. Before laying out this argument in detail, let us first review Kuhn's account of theory choice.

2.5.1   *Kuhnian Theory Choice and the Idea of Virtue Convergence*

In his *Structure of Scientific Revolutions* (1962/1996), T. S. Kuhn claimed that paradigm change, such as the change from Newtonian to relativistic mechanics, or from the phlogiston to the oxygen theory of combustion, 'cannot be . . . forced by logic [or] neutral experience' (149). Rather, each paradigm comes with its own set of evaluation criteria. Whenever scientists have to choose between paradigms, 'each paradigm will be shown to satisfy more or less the criteria that it dictates for itself and to fall short of a few of those dictated by its opponent' (109). In other words, paradigm change is circular in the sense that changing a paradigm must rely on the evaluation criteria that the new paradigm identifies as important (which will be different from the criteria identified as important by the old paradigm). About 10 years after *Structure*, in a seminal paper on theory choice, Kuhn (1977a) tried to answer those who (rightly) accused him of putting the case for relativism. Departing from *Structure* to a degree that he probably did not quite realize, Kuhn in this paper advanced the view that there is a universal set of theoretical virtues on the basis of which scientists assess theories. Kuhn, without claiming either originality or completeness, mentions five prominent virtues: empirical accuracy, (internal and external) consistency, scope, simplicity, and fertility.[16]

Kuhn – slightly reluctantly – distinguished between objective and subjective elements of theory choice (325). The former concerns the set of virtues involved in theory choice. Kuhn writes, 'the criteria or values deployed in theory choice [i.e., the virtues] are fixed once and for all,

---

[16]   Theoretical virtues are also sometimes denoted as 'values'. In fact, Kuhn himself suggested that label. I prefer 'virtues' because 'values' have ethical connotations. Recently there has been a debate about the virtues of *scientists* making theory choice (Stump 2007; Ivanova 2010). My discussion, instead, focuses on the virtues of *theories*.

unaffected by their participation in transitions from one theory to another' (Kuhn 1977a, 335). Although Kuhn assigned universality to the five standard theoretical virtues, he believed that there was a lot of room for *legitimate* disagreement among practitioners in deciding which theory to adopt. Each scientist, Kuhn claims, has different weighting preferences concerning the standard theory-choice criteria. Whereas some prefer simpler theories, for instance, others prefer more unified theories, and so on.[17] It is this subjective element that led Kuhn to the conclusion that there is 'no neutral algorithm' for theory choice to which all practitioners would be bound (Kuhn 1962/1996, 199). Thus, 'two men fully committed to the same list of criteria for choice may nevertheless reach different conclusions' (Kuhn 1977a, 324). The subjective element of theory choice, however, does not imply that theory choice would be arbitrary. The standard criteria of theory choice are not projections; they map onto actual theory properties. Theories really are accurate, consistent, fertile, and so on, or they are not.[18]

Another remark by Kuhn seems to undermine the objective element in theory choice: theoretical virtues are 'imprecise' or 'ambiguous'. By this he meant that different practitioners might refer to different properties of a theory with the same term. For example, one practitioner might refer to quantitative parsimony and another to qualitative parsimony when calling the theory 'simple'. As a further example, one practitioner might judge the Copernican system of the planets simpler because it represents the retrogressive motion in simple terms, and another practitioner might judge the Ptolemaic system as simple as the Copernican system because the Copernican system, too, made use of a large number of epicycles (Kuhn 1957).[19] Yet this problem should constitute no major obstacle for theory choice: barring the much-criticized Kuhnian communication failures, practitioners should be able to specify to their peers what properties they are referring to. Practitioners might then still disagree about how these two kinds of simplicity ought to be weighted, of course. The ambiguity problem thus arguably reduces to the weighting problem (Okasha 2011).[20]

---

[17] Although Kuhn thought the five standard virtues relevant to theory choice throughout the history of science, he thought that the 'application of these values' and the 'relative weights attached to them' changed (335).

[18] This can be gleaned from the fact that Kuhn insisted that theory-choice decisions are judgements, not matters of pure taste (Kuhn 1977a, 336f.).

[19] The precise number of epicycles used depends on which version of the Ptolemaic system one compares to the Copernican system. For an informative discussion, see Palter (1970).

[20] Some readers suggested to me that Kuhn believed that the virtues were *intrinsically* and *inextricably* vague, with no disambiguation being possible. I have found no evidence in Kuhn's text for this suggestion. Regardless, disambiguation of theoretical virtues may not always be unequivocal, of

A similar point can be made with regard to empirical accuracy. In a theory-choice situation, one theory might be empirically accurate with regard to one set of evidence, and another theory might be empirically accurate with regard to another set of evidence. In such a case, we would of course have to fine-grain the virtue of empirical accuracy. Scientists may then disagree as to whether one or the other data set is to be given preference when it comes to the choice between the two theories.

The weighting problem is contingent on another part of the Kuhnian picture of theory choice. According to Kuhn, as a matter of empirical fact, the virtues 'repeatedly prove to conflict with one other' (322). In other words, as a matter of empirical fact, theories repeatedly do better than others with regard to some criteria but worse with regard to others. When there is conflict, and when scientists have different weighting preferences, there will be diverging theory choices. For convenience, let us refer to the claim that the virtues are often in conflict with each other as the *conflict thesis*, and instances of theory-choice situations with conflicting virtues simply as *conflict*.

Conflict can also occur within the category of one particular virtue. Take for example empirical accuracy. One theory may perform better with regard to one set of data, and another theory better with regard to another set of data. Scientists with different preferences will end up choosing different theories. This is another reason for why Kuhn thinks theory choice is often indeterminate (322f.). Again, the 'subjective' element of theory choice – i.e., the interests and preferences of the investigator – will influence which data set an individual will assign greater weight and accordingly which theory she will end up choosing.

Interestingly, and somewhat counter-intuitively, the conflict thesis can explain theory-choice convergence. Again, the conflict thesis is an empirical thesis. That is, at least prima facie, there is nothing intrinsic in the virtues or their relationships that would cause conflict. Therefore, there should be situations in which there are theories that do better than any other theory with regard to all virtues. In that case, the subjective element of theory choice, which Kuhn was so keen to stress, simply cancels out: If I prefer simple theories and you prefer unified theories, then we will adopt different theories when there is no theory that has both of these properties. But when there is a theory that is *both* simple and unified and its

course. There may be boundary cases. But boundary cases do not necessarily imply that we cannot reach agreement. In our everyday life, for example, we manage pretty successfully to agree on what we consider to be a bald person, despite baldness being a standard example of a vague predicate.

competitors are not, then we will end up choosing the same theory *despite* our diverging preferences.[21]

Perhaps even more interestingly, the conflict thesis offers a new argument for realism. Roughly, it goes like this: if the conflict thesis is true and there are only sometimes theories that exhibit several or even all of the standard virtues, then it would be a strange coincidence if a theory had all the five virtues, was embraced by all scientists, and was not true. It is easy to see: just like the standard NMA for realism, this is a no-coincidence argument. I will therefore refer to it as the no-virtue-coincidence (NVC) argument. And yet, as I will argue, NVC is more powerful than the NMA: it offers an answer to Magnus and Callender's challenge.

Before proceeding, however, let us note that the conflict thesis does seem to possess a good deal of prior plausibility. If it were easy to construct theories that possessed all of the standard virtues, and most of the theories we could come up with possessed all of the standard virtues, then a theory possessing all of the virtues wouldn't warrant being singled out as a hopeful truth-candidate. At the same time, scientists would have a hard time making their theory choices when faced with a range of theories that would all score high on all dimensions of theory choice. But very often scientists do come to an agreement as to what theory to embrace. Although the conflict thesis thus does seem plausible, it is of course an open question *how* frequent it is that only a few theories or only one theory at a time possesses all the standard virtues. This question is an *empirical* one and beyond the scope of this book.[22]

2.5.2 Earman Convergence

Although clearly related, the NVC differs from the NMA in that it appeals to the persuasive power of the convergence of several independent information sources. Whereas the NMA exploits the fact that there are so many ways in which a theory could have been wrong, the class of arguments of which the NVC is a member banks on the fact there are so many ways in which each information source could have produced a result inconsistent with the other

---

[21] Kuhn, in his seminal article on theory choice, firmly focused his attention on disagreement. All he did write about agreement was that 'much work, both theoretical and experimental, is ordinarily required before the new theory can display sufficient accuracy and scope to generate widespread conviction' (332).

[22] Note that Okasha (2011) also implicitly assumes the conflict thesis. His impossibility theorem does not apply, though, to cases in which we are interested here, namely cases in which a theory is virtuous along all five dimensions and its competitors are not. In those cases, there is a unique algorithm for theory choice.

sources. Several philosophers have used related 'convergence' arguments. Arguing against the thesis of theory-ladenness of observation, for instance, Hacking (1983) pointed out that it would be a strange coincidence if several of our instruments (e.g., the light and the electron microscope), presupposing different background theories, were to produce the same data if the data were not correct. Likewise, Salmon (1984) pointed out that it would be an inexplicable coincidence if J.-B. Perrin's half-dozen experiments in 1911 had all produced the same value of Avogadro's number, and that number had not been correct (see also Cartwright 1983; van Fraassen 2009; Chalmers 2011; Psillos 2011).[23] These 'convergence' arguments are in fact analogous to arguments for the trustworthiness of witness reports in the case of several independent witnesses reporting the same murderer: we'd be compelled to believe that several witnesses tell the truth if they independently of each other report the same murderer (i.e., without coordinating their beliefs) – even when the individual reliability of the witnesses is poor. Lewis (1946) concludes that 'this agreement [between witness reports] is highly unlikely; the story any one false witness might tell being one out of so very large a number of equally possible choices' (246). In other words, the probability that there is convergence in the witness reports makes it unlikely that the witness reports are unreliable. Here I will argue, analogously, that a theory possessing all of the Kuhnian virtues and being judged true on the basis of its virtues by the scientific community makes it likely to be true.

The intuitive persuasive power of convergent witness reports can be made precise by employing Bayes' theorem (Earman 2000). Let $P(V_i|T)$ represent the probability that a witness i gives a report V that a crime T happened when that crime actually happened, $P(V_i|\neg T)$ the probability that a witness reports a crime when the crime did not happen, and $P(T)$ the prior probability of the crime itself.[24] Assuming that the witnesses are equally reliable and independent, the following equalities hold:

$$P(V_1 \cap \ldots \cap V_n) = P(V_1)P(V_2)\ldots P(V_n) = P(V)^n$$

$$P(V_1 \cap \ldots \cap V_n|T) = P(V_1|T)P(V_2|T)\ldots P(V_n|T) = P(V|T)^n$$

$$P(V_1 \cap \ldots \cap V_n|\neg T) = P(V_1|\neg T)P(V_2|\neg T)\ldots P(V_n|\neg T) = P(V|\neg T)^n.$$

---

[23] For a more general discussion of robustness arguments in science, see Hudson (2013).

[24] Earman also conditionalizes on the witnesses' background knowledge and evidence, which I'll leave out here for the sake of simplicity. Earman develops the argument with regard to the occurrence of miracles instead of crimes.

The posterior probability of the truth of a report given n witnesses making the same observations, by Bayes' theorem, is

$$P(T|V^n) = \cfrac{1}{1 + \left[\frac{1-P(T)}{P(T)}\right] \cdot \left[\frac{P(V_1|\neg T)\cdot P(V_2|\neg T)\cdot...\cdot P(V_n|\neg T)}{P(V_1|T)\cdot P(V_2|T)\cdot...\cdot P(V_n|T)}\right]}$$

which, assuming equally reliable witnesses, reduces to

$$P(T|V^n) = \cfrac{1}{1 + \left[\frac{1-P(T)}{P(T)}\right] \left[\frac{P(V_i|\neg T)}{P(V_i|T)}\right]^n}$$

Let us refer to this equation as the *Earman convergence equation*.

Earman points out that for the occurrence of a crime to be likely, given all the witness reports, the witnesses need not be reliable in the absolute sense. All that is required, rather, is that the witness reports be reliable in the relative sense, so that $P(V_i|\neg T) < P(V_i|T)$, since in that case as n → ∞, $\left[\frac{P(V_i|\neg T)}{P(V_i|T)}\right]^n$ → 0, and $P(T|V^n)$ → 1, *regardless* of how low $P(T)$ is.

2.5.3   Converging Virtue Judgements

In the context of theory choice, I suggest we interpret the foregoing probabilities in the following way. Let $P(V_i)$ stand for the probability of a scientist i deeming a theory T true on the basis of some virtue V of T, and let $P(T)$ stand for the probability of T being true. Let $P(V_i|T)$ then be the conditional probability that T would be correctly judged true on the basis of V by scientist i, and $P(V_i|\neg T)$ the conditional probability of T being incorrectly judged true on the basis of V by scientist i. $P(T|V^n)$ is then the posterior probability of T being true given that it was judged true on the basis of V by n scientists. If we now take into consideration that theories can be virtuous along different dimensions, then

$$P\left(T|(E \cdot C \cdot S \cdot U \cdot F)^n\right) = \cfrac{1}{1 + \left[\frac{1-P(T)}{P(T)}\right] \left[\frac{P(E_i\ C_i\ S_i\ U_i\ F_i|\neg T)}{P(E_i\ C_i\ S_i\ U_i\ F_i|T)}\right]^n}$$

where E_i, C_i, S_i, U_i, F_i is a scientist i's judgement about a theory being true based on that theory's empirical accuracy, consistency, simplicity, and fertility, respectively, in line with the Kuhnian framework of theory choice. The posterior probability (the left-hand side of the equation) thus measures the probability that a very virtuous theory (i.e., a theory with all five virtues) is true, given that it is being judged true by n scientists on the basis of that theory possessing all five virtues.

Note that just like in the case of the witnesses in Earman's equation, we are assuming that scientists' judgements about the theory's truth on the basis of its virtues are fully independent. This is consistent with Kuhn's view that scientists' theory-choice preferences are subjective and diverse (cf. Section 2.5.1). We also assume that the virtues themselves are independent: the fact that a theory possesses one virtue will not make it more (or less) likely that it possesses another virtue. We'll get to that in more detail in a moment, but regardless of whether the virtues are independent, so long as it is the case that *scientists* are independent and relatively reliable – i.e., more likely to judge a theory true on the basis of its virtues when it's actually true – it will still be the case that when n (i.e., the number of scientists) $\rightarrow \infty$, the posterior probability $P(T|V^n) \rightarrow 1$. With regard to Magnus and Callender's challenge, this means that the base rate of true theories need not be high for no-miracles arguments like the NVC to have any traction. What's more, if n really were to converge to infinity, the base rates could be ignored without NVC being undermined.

Of course, in any realistic scenario, the number of scientists n will not converge to infinity. Thus $P(T)$ cannot be arbitrarily small. Yet $P(T)$ may still be so small that the realist may argue that despite our ignorance about the precise value of the base rates, the chances are good that a very virtuous theory is true, if only a low $P(T)$ is granted by the antirealist. It will not win the realist the argument against a very hard-headed antirealist, but it will make the realist's argument much more unassuming. And the antirealist must grant *some* small value for $P(T)$; otherwise, the Bayesian formalism is simply ill-defined. We will consider further objections in Section 2.5.5.

NVC depends on the Kuhnian framework of theory choice in the following way. As we noted previously (Section 2.5.1), when different theories have different virtues and scientists different virtue preferences, then scientists will end up choosing different theories; some might deem true theories that are simple (but not fertile), others might deem those theories true that are fertile (but not simple), etc. On the other hand, if a theory does possess all the virtues, scientists will choose the *same* theory *despite* different virtue preferences. Now, in our formalism we assume that there is a theory that does possess all the virtues. Accordingly, the formalism represents scientists' judgements as the joint probability that scientists judge a theory to be true on the basis of all of the theory's virtues. And this makes good sense: after all, it is reasonable to assume that scientists judge a theory not only on the basis of the virtue they prioritize, but also on the basis of the other virtues. In a more realistic representation of theory choice than the one provided here, the different virtue judgements would have to be weighted. But since this would complicate matters unduly, we shall refrain from incorporating weights. It suffices for our purposes that weights would make a real difference in theory-choice situations in which there is no theory that possesses all the virtues – situations, that is, which we are not interested in here.

As mentioned earlier, we also presume here that scientists' judgements about a theory's truth on the basis of that theory's virtues are independent. This is consistent with Kuhn's view that scientists' theory choice preferences are subjective and diverse. Another question is whether the virtues themselves are independent. That is, is it plausible that a theory's possessing one virtue does not raise or lower the probability of it possessing another virtue? I think it is. Consider for example simplicity, fertility, unifying power, and consistency in relation to empirical accuracy. It seems obvious that a theory's internal consistency doesn't imply anything about empirical accuracy (and vice versa), as the latter is a relation between the theory and the world, whereas consistency are theory-internal relations. Many philosophers believe something similar about simplicity (cf. Chapter 1). The relation between empirical accuracy and the other virtues is perhaps less obvious, but I think there is a good case to be made for their independence as well. Let's start with fertility. If a theory's fertility and empirical accuracy were dependent, then we could infer one from the other. But that's not the case. A theory being empirically accurate doesn't tell us whether or not the theory has any novel success (i.e., the standard form of fertility); in fact, many theories in the history of science were empirically accurate but had no novel success. Conversely, a theory having novel success with regard to some phenomena doesn't tell us whether the theory is empirically accurate: there may be phenomena (other than the one the theory successfully predicted) which the theory does not manage to accommodate. Or take unifying power. A theory accommodating a large number of phenomena doesn't tell us whether the theory has unifying power: the theory might just be an incoherent conjunction of very narrow hypotheses. Conversely, the fact that a theory has unifying power does not imply that it is empirically accurate (e.g., string theory unifies the four fundamental forces of nature but is not even testable). Lastly, because a theory is empirically accurate with regard to the empirical consequences that can be derived from it doesn't imply that it is actually consistent with other (empirically accurate) theories. The inconsistency of quantum mechanics and general relativity is a prominent example.

Overall, then, there are good grounds for thinking that most virtues are independent.[25] The assumption that the virtues are independent will be required for estimating the 'error rates' of scientists' judgements for each virtue separately, which we will attempt in Section 2.5.4. We should stress again, however, that the question of whether or not the virtues are independent does not affect the NVC itself, as $P(T|V^n) \to 1$ when n $\to \infty$, even when the virtues are dependent; all that is required is that the scientists' judgements are independent and relatively reliable (in the sense specified previously).

2.5.4 Estimating the Error Rates

As we saw in Section 2.5.3, for the NVC to go through, scientists' virtue judgements need not be absolutely reliable. They only need to be relatively reliable; that is, the true positive rate $P(V_i|T)$ must only be larger than the false positive rate $P(V_i|\neg T)$. Is that the case?

In order to address this question, we shall exploit an interesting inverse relationship between the 'error rates': a high false positive rate implies a low true negative rate and vice versa since $P(V_i|\neg T) = 1 - P(\neg V_i|\neg T)$, and a high true positive rate implies a low false negative rate and vice versa since $P(V_i|T)) = 1 - P(\neg V_i|T)$.[26] We shall interpret the true negative rate $P(\neg V_i|\neg T)$ as the probability of a scientist i to *correctly* judge a theory T to be *false* on the basis of it not being virtuous, and the false negative rate $P(\neg V_i|T)$ as the probability of scientist i to *incorrectly* judge *false* a theory T on the basis of it not being virtuous. In what follows, we shall use these relationships to estimate the error rates.

Although the probabilities scientists assign to a theory based on its virtues will of course differ from one individual to another (in agreement with the Kuhnian picture of theory choice), an individual scientist judging a theory likely to be true when there is no good reason to (and vice versa) would amount to this scientist being irrational. The error rates can thus be viewed as constraints on rational theory-choice decisions.

First consider empirical accuracy. If a theory is not empirically accurate, it presumably is no candidate for being true: empirical accuracy is a necessary condition for truth. So if a theory is true, it would have to be empirically accurate and accordingly be judged empirically accurate by

---

[25] One may argue that simplicity and unification, for example, are indeed related: the more phenomena a theory accommodates on the basis of fewer principles or entities, the more unifying the theory should be (cf. Chapter 1).

[26] $P(V_i|T)$ is also known as the sensitivity, and $P(\neg V_i|\neg T)$ as the specificity of a test.

scientists. Thus $P(E_i|T)$ must be 1.[27] On the other hand, empirically accurate theories may of course simply save the phenomena without being true. Thus, $P(E_i|\neg T)$, the false positive rate, *could* in principle also be 1. However, as a matter of fact, it cannot, for if $P(E_i|\neg T) = 1$, then $P(\neg E_i|\neg T)$; i.e., the probability of a theory being correctly judged false on the basis of it not being empirically accurate would be zero, because $P(\neg E_i|\neg T) = 1 - P(E_i|\neg T)$, as mentioned before. But that is extremely implausible. Indeed, a lack of empirical accuracy is probably the best criterion scientists can go by when judging a theory to be false. Thus, $P(\neg E_i|\neg T)$ must have some positive value and $P(E_i|\neg T)$, accordingly, a value lower than 1. Since, as determined earlier, $P(E_i|T) = 1$ and $P(E_i|\neg T) < 1$, the latter is lower than the former and we can therefore conclude that empirical accuracy is indeed relatively truth-conducive. That means that although empirical accuracy is by no means a guarantee for truth (after all, we're dealing here with probabilities), and although empirical accuracy must not make it even likely that a theory is true (i.e., it must not even be the case that $P(E_i|T) > .5$, although here it appears to be), it is still more likely that a theory is empirically accurate when it is true than when it is false. Importantly, this does not necessarily mean the converse – namely that a theory which is empirically accurate would be likely to be true. This probability is the probability of the left-hand side of our Earman equation derived in Section 2.5.2, namely $P(T|V^n)$, and is not to be confused with the error rate $P(V_i|T)$, in this case $P(E_i|T)$.

Next we discuss simplicity. Van Fraassen (1980) and others have argued that a theory's being simple gives us no grounds for thinking that the theory is true or likely to be true. Although this is generally accepted, I have provided reasons for the contrary view in Chapter 1. If we were to assume that there were indeed no good grounds for simplicity being truth-indicative, it would *not* be the case that $P(S_i|T) > P(S_i|\neg T)$, i.e., that the probability of a scientist i correctly judging theory T to be true on the basis of it being simple could be larger than the probability of i *falsely* judging theory T to be true on the basis of it being simple. As we argued in the first chapter of this book, however, van Fraassen's arguments against simplicity being epistemic are flawed (because they attack ontological assumptions about the world rather than any epistemic rationale for simplicity

---

[27] The empirical accuracy of a theory may not be absolute: the theory may be empirically accurate with regard to only some (but not other) relevant phenomena. As explained in Section 2.5.1, we can subdivide the virtue of empirical accuracy into empirical accuracy with regard to one set of phenomena and empirical accuracy with regard to another set of phenomena.

considerations). We also offered the explanatory-evidential rationale for why simplicity considerations are genuinely epistemic. We may therefore very well assume that $P(S_i|T) > P(S_i|\neg T)$, whilst keeping in mind that simplicity considerations are context-dependent: the relation should hold in all contexts other than those where we are faced with theories that postulate few, but not *enough*, entities or principles for explaining the phenomena (cf. Chapter 1).

Let us now turn to unifying power. Theories that have been singled out as approximately true by the realists generally manage to unify the phenomena. Einstein's theory of relativity, the standard model of particle physics, the modern synthesis in evolutionary biology, plate tectonics, etc., are cases in point. Thus a true theory, for the realist, is likely to unify, and therefore $P(U_i|T)$ – i.e., the probability of a theory being correctly judged true on the basis of it unifying the phenomena should approach 1. Of course, the antirealist would reject not only the idea that we can ever know whether a theory is true, but also the conditional claim that *if* a theory were true, it would be likely to unify (see Chapter 1). Indeed, since the antirealist believes that unifying power is a *purely* pragmatic virtue (cf. Chapter 6), she would insist that $P(U_i|T)$ approaches 0. But that would imply that it is likely that true theories (whose existence the antirealist does not in principle deny) will reliably *not* be recognized as true by virtue of their unifying power, which would be rather strange indeed. Scientists clearly do recognize when theories accomplish unifying feats *and* take such feats as reasons to believe in them (see Chapter 6). The antirealist can of course ignore this fact in principle, but not if she wants to take naturalism seriously, as any philosopher of science should (see Chapter 7). Furthermore, it seems glaringly obvious that false theories have only rarely, if ever, achieved the kinds of unification that have regularly been accomplished by theories such as Newtonian mechanics, Darwin's theory of evolution, etc. But that means that the false positive rate, $P(U_i|\neg T)$, should be rather low. Hence, overall it seems justifiable to assume that $P(U_i|\neg T) < P(U_i|T)$, as required by the *Earman equation*.[28]

Consider consistency next. Kuhn lumps together internal and external consistency. And yet they are best treated separately. Let's start with internal consistency. Clearly, a true theory should be internally consistent (for short, C_{I_i}), i.e., $P(C_{I_i}|T)$ should be 1. It follows that the false negative

---

[28] Again, if the unifying power of a theory is related to its simplicity, then unification and simplicity could not count as two independent virtues. My argument could thus be based on only one of these virtues. On which would depend on which virtue is considered more fundamental. I will leave this question open here.

rate should be o.[29] The false positive rate $P(C_{I_i}|\neg T)$ is much more difficult to assess. There are probably indefinitely many false theories out there that are internally consistent, which would make $P(\neg C_{I_i}|\neg T)$ approach 1. Yet there will also be at least as many false theories that are not even consistent. We also know that $P(\neg C_{I_i}|\neg T) = 1 - P(C_{I_i}|\neg T)$, so $P(\neg C_{I_i}|\neg T)$ and $P(C_{I_i}|\neg T)$ cannot both be equal to 1 or close to 1. In the face of our ignorance about the precise values, it seems most reasonable to set both expressions to 0.5. Since, as mentioned earlier, $P(C_{I_i}|T)$ must be 1, that would give us $P(C_{I_i}|\neg T) < P(C_{I_i}|T)$ in conformity with the requirement of the Earman equation.

Now consider external consistency (C_{E_i}). True theories, many realists believe, must be consistent with our background knowledge (Boyd 1983; Lipton 1991/2004; Psillos 1999). Thus $P(C_{E_i}|T) = 1$, and accordingly the false negative rate $P(\neg C_{E_i}|T)$ would have to be o. What about the false positive rate $P(C_{E_i}|\neg T)$, i.e., the probability that a theory is judged true on the basis of it being externally consistent if it is actually false? Although there are probably many false theories that are consistent with our background knowledge, there must be a much larger amount of false theories that are inconsistent with our background knowledge, which would mean that $P(\neg C_{E_i}|\neg T)$ should approach 1. But since $P(C_{E_i}|\neg T) = 1 - P(\neg C_{E_i}|\neg T)$, $P(C_{E_i}|\neg T)$ would then approach o. In that case, our condition of $P(C_{E_i}|\neg T) < P(C_{E_i}|T)$ seems to be well satisfied.

Lastly, consider fertility. As mentioned previously, fertility is standardly construed in terms of novel success. And the capacity to successfully predict novel phenomena, as we have already noted, is indeed the virtue most cherished by the realist. In fact, it is often treated as a necessary condition for a theory being true.[30] Thus $P(F_i|T)$ should be equal to 1. Conversely, realists consider false theories very unlikely to generate successful novel predictions. $P(F_i|\neg T)$, thus, ought to be close to o. The condition for our Earman equation is again well satisfied. If the antirealist wishes to challenge this, she would have to present evidence for false theories being capable of producing novel success. Magnus and Callender are right to stress that a handful of cases where false theories allegedly produced novel success won't suffice to challenge the realist. Not

---

[29] As discussed in Chapter 1, there may be some cases of genuinely inconsistent theories in science. But even if there really are, they seem to be very rare.

[30] As I shall explain in Chapters 3 and 4, I'm more relaxed about this requirement for realist commitment.

only need there be a more substantial data base, but, as we noted previously, the antirealist must make a case not only for false theories producing novel success, but also for a theory's false posits doing so (Psillos 1999). Whether or not this has been achieved *in even just a single case* is still up for debate (cf. Section 2.1). In any case, the foregoing reasoning allows us to set $P(F_i|\neg T)$ to a low level. And once again, this gives us $P(F_i|\neg T) < P(F_i|T)$, in agreement with our requirement.

In sum, then, it appears that for almost all theoretical virtues considered here (maybe even for simplicity), $P(V_i|\neg T) < P(V_i|T)$ holds and that therefore our NVC, based on the *Earman convergence equation*, against the antirealist succeeds: for n $\rightarrow \infty$, $\left[\frac{P(V_i|\neg T)}{P(V_i|T)}\right]^n \rightarrow 0$, and therefore $P(T|V^n) \rightarrow$ 1, regardless of the value of $P(T)$. Less formally, since the false positive rate is smaller than the true positive rate, the more scientists embrace a theory on the basis of its virtues, the more likely it is that it is true, even when the base rates of true theories is very small. As $P(V_i|T)$ need not be > 0.5 for this argument to go through, this means that the realist can assume that the theoretical virtues are only *minimally* truth-conducive (so that $P(V_i|\neg T) < P(V_i|T)$). In fact, we *argued* in this section that the theoretical virtues are indeed truth-conducive in this minimal sense.

2.5.5   *Objections and Clarifications*

Although we sought to set the error rates with a great deal of charity (consider, e.g., simplicity), the antirealist might nevertheless have specific objections to how we set them. Such objections are not necessarily bad news for the purposes of this chapter. On the contrary, as this chapter sought to undermine Magnus and Callender's claim that the realism debate is irrational, pointed objections against the way the error rates are set would actually support the conclusions cited here, as they could be evidence for a rational debate. And although it might be over-optimistic to expect that the realist and the antirealist will come to any agreements (as is usually the case in philosophy), arguments about the error rates would be decidedly more tractable than arguments about the base rates.

It is worth recalling that we have already established that the antirealist cannot undermine the NVC by simply setting the base rates to 0: that would render the formalism ill-defined. Yet it is widely accepted that the NMA is suitably expressed in Bayesian terms. There is another reason why the antirealist cannot help herself to such an escape: the antirealist is a sceptic, not a dogmatist. That is, antirealists think we have no good reason to believe that our theories are true – not that we have good reason to believe that our theories are false. But setting the base rates to 0 would mean insisting that virtuous theories are *bound to be false*.

There is a more severe problem for the argument presented in this chapter. As mentioned in our discussion, the number of scientists n can of course not go to infinity; the size of the scientific community is limited at any given time. Accordingly, there will be values for the base rate $P(T)$ for which it will no longer be the case that it is likely that a theory is true when very virtuous; i.e., it will no longer be the case that $P(T|V^n) > 0.5$ – as the realist would like to have it. The antirealist could thus simply adopt values for $P(T)$ which prevent the NVC from going through. It would seem that Magnus and Callender's observation would then still hold. At the very least, however, such a move by the antirealist would be utterly ad hoc, if not question-begging: the antirealist would have no other motivation for setting her base rates than to avoid the realist's argument of succeeding. The antirealist would thus damage the possibility of a rational debate about the error rates, when such debate is clearly available. If the antirealist wishes to retain the possibility of a rational debate (which she of course should), then, to repeat, she would have to argue convincingly that the false positive rate is larger than the true positive rate. In that case – and presuming that the theory in question will be embraced by enough scientists – the prior $P(T)$ becomes negligible.

Although the number of scientists n is not going to converge to infinity, it is still the case on my account that the more scientists embrace a theory, the more likely it is to be correct. One might consider this to be counterintuitive. However, I do believe it is plausible that a theory that is subjected *to more scrutiny* when assessed by a bigger scientific community is less likely to be false when accepted as true by capable scientists: the probability that the scientists in questions are mistaken (e.g., because they are mistaken about the empirical support the theory enjoys) is simply lower the more critical eyes there are.

Although I do claim that virtue convergence enables scientists with diverse virtue preferences to choose the same theory and that such virtue convergence gives us good reason to believe that the theory in question is true, I must emphasize that I do not claim that virtue convergence is necessary for realist commitments. In other words, I wouldn't want to deny that a realist attitude may be warranted also in cases in which there is no virtue convergence. For example, there could be cases where the vast majority of scientists chooses a theory T_1 possessing a single virtue V_1

instead of another theory T_2 that possesses several virtues – say, V_2, V_3, and V_4. My formalism could not accommodate such a case, because it presumes that the theory under consideration possesses all the virtues. In fact, I'm not sure such a case ought to be accommodated by my formalism, as it is not quite compatible with the Kuhnian picture of theory choice according to which scientists' theory-choice preferences are diverse; this rules out situations in which a majority of scientists *jointly* spurns several virtues in favour of the *same* single virtue. Even regardless of the Kuhnian picture of theory choice, it just seems odd that scientists would embrace a theory with just a single virtue rather than one with several. At any rate, I would not want to deny that there may be *other* good grounds for taking a realist attitude towards theories. It is thus no part of my picture that virtue convergence is necessary for realist commitments.

One may fear that the argument proposed here on the basis of the Kuhnian framework of theory choice invites relativism. My argument suggests that we be realists about theories that are held to be true by numerous scientists on the basis of the theory's virtues. Does that not subject us to the risk that scientists, at some point in time, may embrace a theory as true that turns out false in the end? More importantly, does that not render realism relative to a social group at a particular time? My answer to both questions is yes. Yet relativism does not follow. First of all, I, like most philosophers, subscribe to fallibilism. That is, we should never think that we possess the ultimately true theory that will remain with us forever. There is always the possibility that nature will teach us better. Since my argument is a probabilistic one, this possibility is always there. A theory embraced by numerous scientists on the basis of it being very virtuous makes it very likely that the theory is true. It doesn't guarantee it, however. Second, although my argument is relative to scientists' judgements, it is so in an unproblematic way. For one, the judgements are, as I explained in Section 2.3, grounded in the *actual* virtues of the theories. So there clearly is an objective basis for these judgements. For another, as I argued in Section 2.5.1, my account puts constraints on the nature of these judgements. It is therefore not the case that *any* judgements could serve as basis for my argument. Rather, the judgements must be rationally constrained ones, in the sense specified in the previous section.

Another objection to the NVC might be to raise doubts about whether the community of scientists would really be diverse enough for the individual scientists to make independent theory-choice decisions. Isn't it in fact the case that on the Kuhnian picture of theory choice, a scientific community is under the strong influence of a paradigm, ensuring concord amongst the scientists' decisions? First of all, let us re-emphasize that the account proposed here is based on Kuhn's picture of theory choice, which can, and in fact should, be treated independently of Kuhn's other views, in particular his (controversial) views about paradigms. But let us bracket off this point for a moment.

In Kuhn's view, scientists are under the influence of a paradigm only in periods of normal science. But theory choice is most relevant in periods of revolution, where paradigms (or paradigm theories) are changed. In these periods, there clearly is diversity of judgements in Kuhn's view. On the other hand, Kuhn in fact believed that *throughout* the development of science, scientists' divergent weighting preferences would go some way to guaranteeing a right balance between conservatism and innovation (Kuhn 1977a, 332–3). Thus, even in periods of agreement on a paradigm, scientists' individual preferences might diverge. Their agreement despite these divergences, as I argued earlier (Section 2.5.1), is explicable when there is virtue convergence.

Let's turn to a different aspect. The NVC assumes that scientists make judgements about the *truth* of a theory on the basis of its virtues. But why should we presume that scientists make such judgements rather than merely acceptance judgements, which do not require any commitment from them regarding the truth of a theory? On the contrary, is it not more plausible that scientists normally do not make such commitments? In fact, I'm happy to say that scientists need not have any commitments about the truth of a theory, and I accept that what I referred to as 'scientists' judgements about the truth of a theory' *can* be construed minimally so that they amount to no more than judgements about empirical adequacy in van Fraassen's sense – that is, a theory's truth concerning observables in the past, present, and future (van Fraassen 1980). Crucially, I want to keep apart the truth in the context of scientists' judgements, i.e., $P(V_i)$, from the truth of the theory $P(T)$. So a scientist might judge a theory only empirically adequate (and not true) on the basis of that theory's virtue, but if her judgement is constrained in such a way that it could be informative about the theory's truth, then her judgements could indeed be evidence for the truth of a theory, namely $P(T|V^n)$. We explored those constraints in our discussion of the error rates where we asked what virtues a theory is likely to have if it were true.

Finally, one might object that scientists' judgements about a theory's truth based on its virtues, contrary to what I assumed earlier, should be sensitive to *degrees* of virtue. That is, in reality, theories are not just simple or not simple, empirically accurate or not, unifying or conjunctive, etc., but rather more or less simple, empirically accurate, etc. However, we can happily admit that our assumption that judgements are binary is indeed an idealization. Reality is regularly more complex than our representations of it. But it does not follow that the argument made here, because of its idealization, must fail for *that reason*. That would likewise undermine large parts of science, in which idealization looms large. To work out the account proposed here for degrees of virtue may still be an interesting project, but it will have to be carried out elsewhere.

2.6   Conclusion

In this chapter I argued, with the help of a point made by Earman in the case of converging witness reports, that a convergence of scientists' truth judgements on the basis of the virtues of a *very virtuous* theory will make it very likely that the theory is true, almost regardless of the value of the base rates. For reasons to do with the finite number of scientists embracing a theory at any particular time, this cannot be a blanket victory for the realist, but it still helps to tilt the balance in the realist's favour. The only avenues left open to the antirealist are now either to cough up, in an ad hoc fashion, base rates that are designed for her to save face, or she can try to argue with us about the error rates. In the latter case, we would have ourselves a rational debate, in spite of what Magnus and Callender feared. In the former case, the antirealist, although sure to win the argument, would win it arbitrarily.

CHAPTER 3

Novel Success and Predictivism

A theory's successful prediction of a new phenomenon, i.e., short novel success, is widely considered amongst philosophers of science, and particularly amongst realists to be the single most important theoretical virtue. In fact, realists generally are committed to the truth of only those theories which manage to produce novel success. The resort to novel success is also one of the realist's standard defences against the PMI, which we discussed in the previous chapter: not only do realists rule out as irrelevant to the realism debate those theories that have failed to produce novel success, but they also apply the divide et impera move to only those theory parts that are central in generating novel success. But is the appeal to novel success justified? In this chapter I shall argue that it isn't. Realist arguments based on novel success should be given up.

Let us follow the usual convention and call *predictivism* the view that predictive success ought to count more in favour of a theory than a theory's accommodation of already-known phenomena and those espousing the view *predictivists*.[1] Predictivism can be traced back to Descartes, Duhem, Leibniz, and in particular Mill and Whewell (cf. Musgrave 1974; Worrall 1985). Popper, too, was a predictivist. Although falsificationism itself does not imply predictivism (it implies only that our theories must be such that they can be proven false by empirical evidence, regardless of whether novel of not), Popper thought that a scientific theory ought to make 'bold' (i.e., novel) predictions and cited Einstein's light-bending prediction as a prime example (Popper 1963/1978). He also made novel success one of his 'three requirements of the growth of knowledge': 'apart from explaining all the explicanda which the new theory was designed to explain, it must have new and testable consequences (preferably consequences of a new kind); it must

---

[1] There is also an extreme form of predictivism according to which *only* novel success, but not accommodative success, counts in favour of a theory. There are only few, if any, philosophers who hold this view. Such a view is, however, entailed by the straightforward interpretation of the widely championed Bayesian confirmation theory (see Section 3.3.2).

lead to the prediction of phenomena which have not so far been observed' (Popper 1963/1978, 241–2). In apparent contradiction with his rejection of induction, Popper also required that novel predictions be confirmed: 'science would stagnate, and lose its empirical character, if we should fail to obtain verifications of new predictions' (244).[2] Popper's student Imre Lakatos, too, thought that novel success was essential to scientific progress; he made novel success the principal criterion for judging whether a research programme – essentially, the developments of a theory – is progressive (Lakatos 1978, 34 and 43; cf. Chapter 4).

One of the most influential versions of the thesis of *predictivism* has been defended by John Worrall in his account of use-novelty. I shall review this account in Section 3.1. In Section 3.2, I will investigate the thesis that the appeal of novel success has something to do with whether or not a parameter in a theory is free, which has also been defended by Worrall and by others. In Section 3.3, I will review approaches to predictivism, which I consider to be misleading. In Section 3.4, I will review important deflationary approaches towards predictivism. In Section 3.5, I discuss one important historical case with bearings on the predictivist thesis, namely the successful prediction of several chemical elements by Dimitri Mendeleev. In Section 3.6, I conclude this chapter with some thoughts on Popperian intuitions regarding predictivism.

3.1 Worrall's Account of Use-Novelty: Weak and Strong

Popper and Lakatos thought of novel success clearly in temporal terms. That is, a theory predicts a new phenomenon when the theory predicts a phenomenon *before* the phenomenon is actually discovered (Lakatos 1970a, 66). And yet one may ask with John Worrall, 'why on earth *should* it matter whether some evidence was discovered before or after the articulation of some theory?' (Worrall 1989b, 148). Instead, Worrall, influenced by the work of Zahar (1973a, 1973b), suggested an alternative construal of novel success:

> **Use-novel success:** A theory T successfully predicts evidence E, iff E wasn't used in the construction of T. If E is use-novel success with regard to T, then E empirically supports T.

---

[2] Popper did not believe that theories could be confirmed, because confirmation presupposes induction, which, Popper accepted, had no justification. Popper's third requirement was simplicity.

The driving idea behind this construal is that a theory T can only be genuinely tested by some evidence E if T was not designed in such a way so as to entail E, because then, the evidence would not constitute a real test for T (Worrall 1989b). Why would it not be a real test? The reason is that 'it is obviously no test of T to ask it to get right some result which had been explicitly incorporated into it in the first place' (Worrall 1989b, 149). When we compare two theories that entail the same evidence where one of the theories had to be modified ad hoc to entail the evidence, then we would clearly prefer the theory that did not have to be modified (Worrall 1989b, 149; 2014). It is therefore 'wrong to regard the downgrading of *ad hoc* explanations and the apparent upgrading of genuine predictions as two separate methodological phenomena'. 'At root', Worrall continues', they are 'the same phenomena' (1989b, 148). Accordingly, we can formulate the inverse of use-novel success as

> **Ad hocness:** A theory T is ad hoc with regard to evidence E if E was used in the construction of T. E lends no or only little empirical support to T.[3]

It is worth mentioning that Worrall sees the use-novelty account, which is also known as the *heuristic account of confirmation*, to specify and to make precise Popper's 'intuitive remarks about testability' (Worrall 1985, 313). Indeed, Popper once wrote that his own requirement of novel success, mentioned earlier, 'seems to me indispensable since without it our new theory might be ad hoc; for it is always possible to produce a theory to fit any given set of explicanda' (Popper 1963/1978, 241–2).

Worrall takes pride in the fact that the use-novelty account seems to be in better accord with the practice of science than is the temporal novelty account. In particular, Worrall believes that his account captures better the fact that sometimes evidence that was not predicted in the temporal sense (but merely accommodated) seems to count *just as much* in the appraisal of theories as evidence that was predicted. Consider, for example, the appraisal of Augustin Fresnel's wave theory of light after its proposal in 1816. Worrall (1989b) discusses in detail the report by a commission for the French Academy of Sciences prize in 1819 that assessed the theory. There was an astonishing prediction that one of the commission members, Siméon Poisson, derived from Fresnel's theory: light shone through a small hole onto an opaque disc should result in a small bright spot at

---

[3] Note that I'm using entailment here rather than logical equivalence (as in the definition of use-novelty), since Worrall may not be committed to the use of evidence in the construction of a theory (which entails that evidence) being the only form of ad hocness.

the centre of the disc's shadow. Contrary to the expectations of some of the commission members, many of whom were Laplacians (including Poisson) and sympathetic to the opposing emission theory of light, this prediction was confirmed. Yet Worrall finds no evidence whatsoever in the report that this astonishing novel success was received with more awe than Fresnel's successful accommodation of straight-edge diffraction patterns, which were known decades before Fresnel proposed his theory. Worrall believes that the use-novelty account can accommodate this historical case (and others) better than perhaps the more intuitive temporal novelty account. Whereas the temporal novelty account seems to *falsely* imply that Fresnel's white-spot prediction should have received higher appraisal from the scientific community than his straight-edge accommodations, the use-novelty account, Worrall claims, implies a symmetry of appraisal when it comes to the straight-edge diffraction patterns and the white spot: none of those phenomena was used in the construction of the theory. Both phenomena were therefore use-novel with regard to Fresnel's theory and rightly counted similarly in the appraisal of Fresnel's theory by the French Academy.

Despite what Worrall says, however, use-novelty does not necessarily imply a symmetry between use-novel accommodations and predictions. Rather, there is reason to think that there is asymmetry even on the use-novelty account: evidence that is not known at the time a theory is constructed *cannot possibly* be used in the construction of that theory. On the other hand, that possibility does exist with evidence that is known when a theory is constructed. Given that scientists may simply not be very explicit about which pieces of evidence they use in the construction of their theories (why should they?), the possibility of constructing a theory based on some evidence or the lack thereof, one could argue, suffices to justify an asymmetry of confirmation between use-novel prediction and accommodation after all. Note also that there exists no possibility of constructing a theory on the basis of the evidence that the theory entails when the theory temporally predicts that evidence. In fact, this is one of the reasons why the proponents of temporal predictivism believe that temporally novel evidence is superior to mere accommodations (Lipton 1991/2004, 140).

There are more general issues with Worrall's account. First of all, one may doubt that the way a theory was constructed ought to impinge on its degree of confirmation. If that were the case, then a theory T entailing evidence E would count less than *the very same theory* entailing *the very same evidence* when a scientist S constructed T on the basis of E, as opposed to when S constructed T without consideration of E. Although this is endorsed by Worrall (1985, 306), others might find this less intuitive. Second, it is not very plausible that the potentially obscure construction process, aforementioned, should have any bearing on the objective assessment of theory–evidence relations. If the construction really mattered to scientists' assessment of their theories, one would expect them to seek out facts about it. But they don't seem to do that. Even if scientists were to do so, could they really assume their peers to always be perfectly honest about their constructions if they knew they could boost the belief of their colleagues in their theories by simply pretending not to have used the evidence which their theories entail?[4]

Worrall's writings in fact contain another, stronger version of use-novelty that is not subject to these worries:[5]

> **Use-novel success:*** A theory T successfully predicts evidence E iff there is no free parameter in T that needs to be fixed on the basis of E in order for T to entail E. Evidence that is use-novel with regard to T empirically supports T.

Conversely, we have

> **Ad hocness:*** A theory T that entails E is ad hoc with respect to E if there is a free parameter in T that must be fixed for T to entail E. Evidence that is not use-novel with regard to T lends no or only little empirical support to T.[6]

It is easy to see why this stronger version of Worrall's is no longer subject to the objections already mentioned. Whether or not a theory possesses free parameters that need to be fixed on the basis of phenomena that the theory entails clearly is an objective matter. It no longer matters who constructed

---

[4] For related criticisms, see Hudson (2007) and Gardner (1982). Leplin (1997) has defended a view of (use-)novelty which is subject to similar objections (Ladyman 1999). See fn. 11. For the view that novelty does not matter at all (in whatever form) and that novelty intuitions can be salvaged by a certain version of test severity, see Mayo (1996).

[5] Worrall – even in his most recent publications on the subject (e.g., Worrall 2014) – does not distinguish between the two versions and presents them as a single account. This has caused a lot of confusion in the literature.

[6] Worrall's account is more complex than my presentation suggests. In particular, Worrall appeals to the condition of independent support: when T is modified in an apparently ad hoc fashion on the basis of E by fixing a parameter for E in T, resulting in T*, and T* makes independent predictions which are supported empirically by E*, then E* also supports T. An apparent ad hoc modification of T is thereby in *some sense* rendered methodologically kosher retrospectively. Yet, for Worrall, modifying a theory on the basis of E so that T can accommodate E is always wrong, so that even when E* supports T, T gets no support from E. I will discuss the independent support condition in Chapter 5. For a more detailed discussion of the intricate details of Worrall's account, see my work (Schindler 2014a).

the theory and how they did so. Whereas on the weak use-novelty criterion it mattered whether a piece of evidence was *actually used* in the construction of a theory, it now matters only whether a piece of evidence was *needed* to construct the theory entailing that evidence. And if a piece of evidence is needed to construct a theory entailing the evidence, then, according to Worrall, this evidence cannot support the theory, because the evidence is 'built into' the theory – i.e., is guaranteed to be accommodated by the theory. With regard to Einstein's theory of relativity and the advance of Mercury's perihelion, which was long known before Einstein proposed his theory, for example, this means that

> there is *no specific parameter* within that *theory that could have been fixed* on the basis of [the Mercury observations] so as to produce a specific theory that entailed those observations. (Worrall 2005, 819)

> [and] it is no part of the heuristic view that it should matter what Einstein was worrying about at the time he produced his theory, *what matters is only whether he needed to use* some result about Mercury in order to tie down some part of his theory. (Worrall 1985, 319)

Note that the strong version of use-novelty implies the epistemic *symmetry* between temporally novel predictions and use-novel accommodations, which Worrall believes the historical facts call for. When we have a theory that contains no free parameter for the phenomena it entails, then these phenomena cannot possibly fail to be use-novel with regard to that theory – regardless of whether those phenomena are predicted by the theory before or after the phenomenon was discovered. Conversely, if a theory does contain free parameters for certain phenomena, then the theory is bound to be non-use-novel with regard to those phenomena, because the phenomena *must* be used in order to render the theory empirically adequate. Interestingly, in that case, there cannot be any temporally novel predictions of evidence pertaining to those parameters, as the parameters cannot be set without having the relevant evidence available.

3.2   Parameter Freedom and Local-Symptomatic Predictivism

Hitchcock and Sober (2004), building on Forster and Sober (1994), too, have appealed to the idea of parameter-freedom in order to motivate their view on novel success. According to them, predictive success is sometimes, but not always, better than accommodative success. This is what makes their predictivism, in their terminology, a *local* form of predictivism. When it is more valuable, it is more valuable because predictive success is indicative or *symptomatic* of another theoretical virtue. Yet when the theoretical virtue in question is apparent, accommodative success will be as good as predictive success. Hitchcock and Sober see themselves as subscribing to the (weak) use-novelty account: 'if we know that a theory was deliberately constructed to accommodate data, we have a prima facie reason to be suspicious of that theory' (5). We have good reason to be suspicious, according to Hitchcock and Sober, because accommodating a set of known data 'run[s] the risk of committing a methodological sin – *overfitting* the data; and overfitting is a sin precisely because it undermines the goal of predictive accuracy' (3). Contrapositively, if we have a theory or model that predicts evidence, then this is evidence for the theory or model not overfitting the data. Hitchcock and Sober use data models, and more specifically polynomials (such as $Y = 4X^3 + 2X^2 + X - 35$) describing data sets, in order to support their view.

Any data set can be fitted to an almost arbitrary degree of precision with some polynomial, whereby a polynomial of degree $(n-1)$ exactly fits n data points (11). Now let us suppose, with Hitchcock and Sober, that we are offered two data models M_1 and M_2 (each of the form $Y = a_n X^n + a_{n-1}X^{n-1} + \ldots + a_1X + a_0$) and a data set D with n data points. M_1 is constructed on the basis of the entire data set D; M_1 fits D almost perfectly. Yet M_1 fits the data at the cost of requiring a polynomial of a high degree (say $n = 10$). M_2, on the other hand, is constructed on the basis of only part of D; call that part D_1. M_2 makes do with a lower-degree polynomial (say $n = 5$). Although M_2 accommodates only D_1, it turns out, it also predicts (in the temporal sense) the other part of D (call it D_2) fairly accurately. That is, when presented with the X-value of a data point, such as the average daily caloric intake, M_2 will fairly accurately predict the Y-value of that data point, such as body weight. Now, even though both M_1 and M_2 fit D, the intuition that Hitchcock and Sober want to elicit is that we should give M_2 more epistemic credit than M_1. Why? Because there is a high chance that M_1, because it is fitted almost perfectly to D, overfits the data; it is likely to have accommodated noise in the data. And noise is 'unlikely to recur in further samples drawn from the same underlying distribution' (11). That is, M_1 invoked a high-degree or 'complex' polynomial for all the wrong reasons. As a general lesson, one should therefore seek to construct data models that hit the right balance between data fit and 'simplicity' of the polynomial (as expressed in its degree). In supporting their case, Hitchcock and Sober appeal to the Akaike Information Criterion, which allegedly provides 'an unbiased estimate of the predictive accuracy of a model' in terms of the number of adjustable parameters of the model (12; see also Forster and Sober 1994). There are other information criteria, such as the Bayesian Information Criterion, but, according to Hitchcock and Sober, all those criteria 'make it abundantly clear why accommodating the data in a particular way – namely, by simply finding a model that fits the data perfectly – is a bad idea if one's goal is to find models that will be predictively accurate' (14). They furthermore emphasize that this insight generalizes to 'any problem in which one is choosing between models that have different numbers of adjustable parameters' (14).

Crucially, Hitchcock and Sober argue that one can reason reversely and take a model's predictive success as an indicator of the absence of over-fitting. With regard to our previous example, the fact that M_2 successfully predicted D_2 can be treated as evidence that M_2 did not overfit the original data set D_2; otherwise, M_2 might not have accurately predicted D_2 in the first place (because it was overfitted to noise). In cases where it is unclear whether or not a model overfits the data, a model's predictive success might therefore be valuable evidence for its 'simplicity', as measured in terms of its number of parameters. Indeed, the risk of overfitting is directly related to the number of parameters: the fewer the parameters, the simpler the model, the less likely the overfit, and the more likely the successful prediction of future data (from the same distribution).

Hitchcock and Sober claim that their account is not limited to data models but extends to more 'high-level theories' such as Fresnel's wave theory of light, mentioned in Section 3.1. Recall Worrall's point about the committee of the French Academy of Sciences not especially highlighting the novel success of Fresnel's theory. Hitchcock and Sober claim that this was so because the committee was impressed by the theory explaining the extant phenomenon of linear diffraction on the basis of just a single free parameter in the theory. They conclude as follows:

> Since Fresnel was able to achieve such an excellent fit with the extant linear diffraction data using a model with only one free parameter, he could not have been overfitting that data. Since the main advantage afforded by the successful prediction of novel data is protection against overfitting, there was no reason to accord special status to the novel prediction in this case. (30)

Yet Hitchcock and Sober do not adduce any positive evidence for the claim that the concern of overfitting really guided the commissioners' reasoning. All Hitchcock and Sober can claim for themselves is that they have established consistency with this case. In fact, it is doubtful whether

Fresnel's theory (or any 'higher level' theory, for that matter) behaves analogously to polynomials, where the paucity of parameters directly translates into high predictive power.

We shall discuss another historical example of theory appraisal in more detail further on, but let us mention two more general points of critique here. First, although Hitchcock and Sober's analysis applies to data of the same distribution, it has been questioned whether it also applies to cases of extrapolation 'where the X values of the new data are beyond the range of X values in the old data', which arguably are the more interesting cases (Lee 2013, 595). Second, Hitchcock and Sober seem to presuppose what is at stake in the debate about novel success: they argue that *if* novel success is a goal in modelling, then perfect fit with the data is a bad thing (3 and 14; see also earlier in this discussion). But again, the predictivism debate is precisely about whether novel success is a goal of science or whether empirical adequacy is the only goal of science. According to that latter view, theories must aim to be consistent not only with the known facts, but also with future observations. However, it is not necessarily the case on that view that theories gain epistemic boosts from making novel predictions successfully. Lastly, it is of course a goal of science not to make our theories accommodate noise. But, contra Hitchcock and Sober, there is no reason why that goal could not be pursued independently of the goal of producing novel success.

3.3 Non-Starters

There are two approaches towards the question of whether novel success constitutes better evidence than accommodated evidence, which I consider non-starters. One approach is motivated through coin-flipping examples, another through Bayesian confirmation theory. Let us consider those in turn.

3.3.1 *Novel Success and Coin-Flipping*

Suppose a coin is tossed 99 times and lands heads and tails approximately half of the time each, in random order. Before the coin is flipped a 100th time, a scientist A, who observed and recorded after the fact the outcomes of all 99 previous tosses, predicts the outcome of the 100th coin toss. Now, consider how much trust we would place in scientist A's prediction as compared with the prediction of another scientist, B, who successfully predicted all 99 previous tosses and who also makes a prediction about the

100th toss. Maher (1988), who advanced the example, suggests that we ought to have greater confidence in B's than in A's prediction about the 100th coin toss. Why? Because scientist B, when correctly predicting the previous 99 coin tosses, demonstrated that she had a 'reliable method' for predicting the outcome of coin tosses. In contrast, scientist A, even if she were to get the outcome of the 100th right, would seem to merely make a lucky guess. Maher takes this scenario to support predictivism.[7]

Various authors have argued that Maher's coin-tossing example is too dissimilar from real scientific examples and therefore has no bearing on whether or not predictivism is true in the scientific realm (Howson and Franklin 1991; Harker 2006; Worrall 2014). One essential dissimilarity concerns the (un-)predictability of events: coin tosses are inherently unpredictable, whereas the phenomena predicted by scientific theories, such as light-bending, are not (Harker 2006). We only trust scientist B more than scientist A in her prediction of the 100th coin toss because we know that she somewhat *mysteriously* managed to successfully predict inherently unpredictable events. Maher's coin-tossing example is also dissimilar to scientific examples in that it's about the wrong thing: methods rather than theories. Maher's example may give us confidence in the superiority of the *method* scientist B uses for making her predictions, but what predictivism in the scientific context is really about is *theories* rather than methods or even the trustworthiness of scientists, as Barnes (2008) has claimed.

Even if we were to grant that Maher's example (or others of that sort) could help us advance the debate about predictivism in *science*, there are doubts that it speaks for predictivism in the first place. Lange (2001) has argued that our intuitions in Maher's example do not speak for the superiority of predictive success per se, but rather for the superiority of unified hypotheses. In order to make this point, Lange 'tweaks' Maher's example and now supposes that the coin tosses produce a strict alternation of heads and tails. In this scenario, Lange contends, the perceived asymmetry between predictive success and accommodative success disappears: whether the sequence of the first 99 coin flips is predicted or just accommodated makes no difference to our confidence in the prediction of the outcome of the 100th toss. The reason we have greater confidence in scientist B in Maher's example, according to Lange, is that, in contrast to scientist A, scientist B seems to have latched onto some non-apparent underlying pattern, as evidenced by her successful

---

[7] White (2003) makes a similar argument for predictivism on the basis of a lottery example.

predictions of the entire sequence of 99 tosses. So what really matters are not so much the predictive capacities of B but rather whether or not the pattern in question has any systematicity. In Lange's tweaked version it is clear that it does, but in Maher's example the hypothesis each scientist forms of the pattern – at least prima facie – must look like an 'arbitrary conjunction' to us. Successful predictions, for Lange, are thus only epistemically superior in cases in which they can give us evidence that the hypotheses formed are not arbitrary conjuncts, as in Maher's example. In other words, Lange too embraces symptomatic predictivism.

Worrall (2014) is not convinced. He proposes a 'tweak of Lange's tweak' to show that scientist B's 100th prediction is indeed more trustworthy than A's. Suppose in Lange's example of a 'tails, heads, tails ...' sequence that A and B make different predictions for the 100th toss: A predicts heads (as to be expected from a simple induction over the sequence, as even tosses should be heads) and B predicts tails. Worrall suggests that in this case we should again trust B's prediction, not A's, because we have seen evidence beforehand of B latching onto the right pattern by correctly predicting (and not just accommodating) all 99 previous tosses. Worrall furthermore criticizes Lange's distinction between arbitrary and non-arbitrary conjuncts as itself quite arbitrary (7): any set of data points can be described by an infinite number of curves, so any of those curves may be considered arbitrary or non-arbitrary with regards to a yet unobserved data point. Science, Worrall concludes, is not about 'pattern recognition', but rather about finding 'underlying explanatory regularities' that can 'reveal hidden order' in the data. Whether or not a pattern appears arbitrary to us thus strongly depends on theories about the data (7–8; cf. McAllister 1997). Worrall believes that Maher's example ultimately supports his use-novelty account, which we have already appropriately criticized.

3.3.2 *Novel Success and Bayesian Confirmation*

Predictivism has also been discussed in the literature on confirmation theory. In particular, straightforward interpretations of Bayesian formalism, by far the most popular formal apparatus used in the confirmation literature, seem to favour predictivism. At the heart of Bayesian confirmation theory is Bayes, theorem, which provides an algorithm for updating one's degrees of beliefs given certain evidence:

$$P(h|e) = \frac{P(e|h) \cdot P(h)}{P(e)}$$

whereby $P(h)$ is the prior probability of the hypothesis in question (e.g., that Einstein's theory of relativity is true); $P(e)$ the probability of an event relevant to the hypothesis (e.g., Mercury's perihelion); $P(e|h)$ the probability of the relevant event occurring if the hypothesis is true (e.g., the probability of Mercury's perihelion conditional on relativity being true); and finally $P(h|e)$ the posterior probability that the hypothesis is true, conditional on the event occurring (e.g., the probability of Einstein's theory of relativity being true, conditional on the probability of Mercury's perihelion).

The Bayesian formalism seems to entail predictivism – in fact, an extreme form of predictivism in which *only* novel evidence can confirm a theory. As Glymour (1980) famously pointed out in what came to be known as the *problem of old evidence*, for evidence that we know, $P(e)$ should be equal to 1. If e is entailed by h, then $P(e|h)$ should be equal to 1 too. But then Bayes' theorem becomes $P(h|e) = P(h)$. That is, the posterior probability equals the prior probability. In our example, the probability that Einstein's theory of relativity is true, conditional on Mercury's perihelion, is equal to whatever our prior probability of Einstein's theory of relativity was before we considered Mercury's perihelion. The evidence has therefore *no effect whatsoever* on our beliefs in the theory in question, which of course cannot be true. In contrast, if e is a novel phenomenon, such as light-bending, the Bayesian formalism would turn into $P(h|e) = \frac{P(h)}{P(e)}$, whereby $P(e)$ would be < 1 and therefore $P(h|e) > P(h)$.

There are various attempts to save the Bayesian formalism despite this prima facie embarrassment. One (not very popular) attempt, made by Howson and Urbach (2006), is to pretend that e, contrary to what is the case, is not known. Counterfactually, one may then ask, How much would e increase my belief in h were I not to know e? This is an approach recently pursued by Steele and Werndl (2013) in their discussion of parameter-fixing in climate modelling. Some, Glymour himself amongst them, have objected to this move on the basis of scepticism regarding counterfactual reasoning. More importantly, perhaps, Howson and Urbach's counterfactual solution negates any difference between actual and just-possible evidence. Actual evidence, treated counterfactually as unknown, may thus boost our confidence in a theory just as much (or even less) than non-actual evidence. But this is of course absurd.

There is another attempt to evade the problem of old evidence, also considered and dismissed by Glymour himself in his original publication. According to this attempt, once we learn that a newly developed theory is supported by the evidence, we learn that the theory entails the evidence (or makes it more likely). The entailment relation is, as it were, a new fact that we now know and that we did not know before the theory was proposed. Thus, we would not conditionalize on e but rather on h *entailing* e. As Howson and Urbach (2006) point out, however, it might well be that we formulate the theory in question with the express goal of explaining e. Then we would know the relevant entailment relationship at the time we inspect e. The old evidence problem would greet us again in simply a slightly different form.

Yet another attempt to escape the problem of old evidence would be for Bayesians to help themselves to Worrall's use-novel conception, according to which, as we saw, evidence known but not used in the construction of the hypothesis in question would count as new evidence. But of course, the use-novelty account – at least in its weak formulation, as we saw in Section 3.1 – faces serious difficulties. Whether the strong form of use-novelty, understood as parameter-fixing, could be successfully put to work in this context I want to leave open. At any rate, what one can reasonably conclude from the very brief review, I think, is that although there might be ways of reconciling Bayesian updating with the practice of science, a significant amount of the work required for doing so will be of a philosophical nature. Also, I believe, Bayesian confirmation theory should reflect scientific practices rather than the other way around. We should not put the cart before the horse and argue that because Bayesianism tells us X, scientists doing non-X are mistaken.[8]

3.4   Deflationary Approaches and Comparative Novel Success

Some authors have argued that novel success has no special epistemic status whatsoever. One of them is the historian Stephen Brush. Brush has investigated several historical cases, with mixed results (Brush 1989, 1994, 1996). Yet with regard to one of his more detailed studies of what is widely regarded one of the most astonishing successful novel predictions, namely,

---

[8] Steele and Werndl (2013) have argued recently that Bayesian confirmation theory can warrant the confirmation of climate models by data which are used to calibrate those models, despite this practice being castigated by climate modellers and despite such 'double-counting' violating Worrall's condition of use-novelty. See Frisch (2015) for a critique. Cf. fn. 17. Steele and Werndl (2016) strike a more conciliatory tone.

the successful prediction of light-bending by Einstein's theory of general relativity (e.g., Popper 1963/1978), Brush's conclusions are clear and in fact somewhat surprising. According to Brush, Einstein's successful prediction of light-bending seems to have had not just as much but in fact *less* weight in physicists' assessment of the general theory of relativity than Einstein's explanation of the advance of Mercury's perihelion, which had been known for decades. Brush offers several reasons why this might be so. First, '[i]n part this was because the observational data [available for Mercury's perihelion] were more accurate – it was very difficult to make good eclipse measurements, even with modern technology' (Brush 1989, 138). Second, 'the Mercury orbit calculation depended on a "deeper" part of the theory' (138). And third, 'physicists recognized that the result [of light-bending] might be explained by more than one theory' (138). In contrast,

> [b]ecause the Mercury discrepancy had been known for several decades, theorists had already had ample opportunity to explain it from Newtonian celestial mechanics and had failed to do so except by making implausible *ad hoc* assumptions. Einstein's success was therefore immediately impressive; it seemed unlikely that another theory would subsequently produce a better alternative explanation. It was a few years before Einstein's supporters could plausibly assert that no other theory could account for light bending, and this phenomenon therefore counted as evidence in favor of Einstein's theory over the others. (Brush 1989, 138)[9]

Achinstein (1994) interprets Brush as defending the view that a theory may receive more confirmation from a phenomenon which it successfully explains than from a phenomenon which it successfully predicts, because in the former, but not in the latter case, scientists have enough time to rule out alternative explanations of the phenomenon. Criticizing this view, Achinstein points out that also in the case of predicted evidence one can call for the consideration of alternative hypotheses before making an ultimate assessment, and one may take as much time as one wishes to consider alternatives, even before any experiment is performed.[10] However, there is reason to think that Achinstein misrepresents Brush's view. His view is not that it is not possible to consider alternative hypotheses and that one cannot, in principle, take as much time as one wishes to consider alternatives when evidence is predicted. Rather, his claim is only that (i) in

---

[9] Brush (1989) also mentions that, in contrast to Mercury's perihelion, it was 'difficult to make good eclipse measurements, even with modern technology' (246).

[10] Achinstein also argues that there can be severe tests for hypotheses even if there is no consideration of alternative hypotheses. See also Mayo (1996).

the case of Mercury's perihelion, astronomers *as a matter of fact* had for a long time failed to find a satisfying explanation before Einstein proposed his idea and that (ii) after the confirmation of light-bending, astronomers needed time to explore alternative explanations to Einstein's before they could conclude that Einstein offered the best explanation of the effect. Brush's claim concerning the asymmetry of explanations and predictive success in favour of the former is thus about a limited period of time and a particular historical setting. Furthermore, Brush seems to claim not *just* that the lack of alternative explanations warranted a higher degree of confirmation of Einstein's successful explanation of Mercury's perihelion. Rather, he seems to say (admittedly without much elaboration) that it is the lack of alternative *non*–ad hoc *explanations* which warranted a confirmatory boost of Einstein's theory. I find this plausible. Incidentally, the lack of alternative non–ad hoc accommodations should probably also be part of the story that needs to be told with regard to light-bending. As Mayo (1996) has pointed out, in response to the light-bending result, several astronomers proposed (Newtonian) alternative explanations, all of which look rather ad hoc. I shall have much more to say about the notion of ad hocness in Chapter 5.

With regard to Einstein's prediction of light-bending, Brush mentions another interesting point made by the physicist Everitt (1980) and Weinberg (1993). The latter makes the same point rather eloquently:

> I think that people emphasize prediction in validating scientific theories because the classic attitude of commentators on science is not to trust the theorist. The fear is that the theorist adjusts his or her theory to fit whatever experimental facts are already known so that for the theory to fit these facts is not a reliable test of the theory. In the case of a true prediction, like Einstein's prediction of light by the sun, it is true that the theorist does not know the experimental result when she develops the theory, but on the other hand the experimentalist does know about the theoretical result when he does the experiment. (Weinberg 1993, 96–7)

I do believe that in the case of Einstein's light-bending prediction, there is indeed more than a suspicion that the experimentalists were guided by Einstein's prediction. More details can be found in Chapter 6.

Brush is not alone in rejecting the evidential superiority of novel success. Harker (2008) argues against any form of predictivism, i.e., not only temporal predictivism but also the use-novel account and symptomatic predictivism. For him, predictivists have not shown that novel success is 'any more reliable at tracking the epistemic virtues of a theory than considerations pertaining to the reputation of the author, advocacy of the expert and so on' (438). Harker also shows sympathy to the concern expressed in the quote by Weinberg, i.e., Harker agrees that the risk of questioning the data when they do not confirm a theoretical prediction is higher for novel phenomena than for phenomena already known: the former are just not as 'thoroughly tested' as the latter (440). Harker believes that the appeal of novel success derives predominantly from cases where theories accomplish the unification of the phenomena, including the successfully predicted phenomenon, without invoking further assumptions. We shall return to this thought also in Chapter 5. Furthermore, in many of the standard historical cases discussed in the literature, Harker believes, the theory which supposedly gains an epistemic boost from novel success explains phenomena that were 'either anomalous or at least unexplained by rival theories' (445). Harker deplores that 'no attention [is drawn by the discussants] to the successful theory's *unique* capacity, among available theories, to explain those phenomena that are considered significant' (445; emphasis added).[11] In particular Harker claims with regard to Fresnel's explanation of straight-edge diffraction that 'at least part of its significance stems surely from the fact that rival theories could not account for those same phenomena' (446; emphasis added).

Given his claim that predictively successful theories are often the only theories that can explain the new phenomena, Harker effectively advocates a different conception of novel success, namely *comparative predictivism*, according to which the question of whether some evidence is novel with regard to some theory depends on that theory's relation to other available theories. This view can be found in Popper (1959a), and it has been defended in detail by Musgrave (1974), Leplin (1997), and Harker (2008, 2013). One can again distinguish between a weak and a strong version. In the weak version, a piece of evidence E is novel with regard to a theory T if there are no theories other than T from which E (or non-E) can be deduced. In other words, only T (and no other theory) says something about E. Conversely, E is not novel with regard to T if there are other theories from which E (or non-E) can be deduced. In the strong version, a piece of evidence E is novel with regard to theory T if there are other theories from which non-E can be deduced. Conversely, E is not novel in that strong version if E can be derived from other available

---

[11] Harker acknowledges his debt to Leplin (1997), who has defended the view that a piece of evidence E is novel with regard to a theory T when (i) T was developed independently of E and when (ii) there were no alternatives to T which explained or predicted E at the time at which T did. The first condition strongly resembles Worrall's weak notion of use-novelty. The second condition has been convincingly disputed by Ladyman (1999). See the following discussion.

theories. Put more simply, in the weak version a piece of evidence is novel with regard to a theory when alternative theories do not predict anything about it, and in the strong version it is only novel when alternative theories predict its absence. It is easy to see that the strong version seems to be unduly strong: if there are no theories other than T that say anything about E, E could not count as novel with regard to T.[12] We should also note that although some proponents of comparative predictivism have defended the radical view that a theory is confirmed *only* by its novel predictions (comparatively construed) and not at all by its accommodations (Musgrave 1974), just like in the case of temporal predictivism, the more reasonable view to take is of course that theories are confirmed by *both* novel and non-novel evidence, albeit more strongly by novel than by non-novel evidence.

Obviously, both the weak and the strong versions of comparative predictivism, contrary to temporal predictivism, are indifferent to the time order of the prediction. Thus, when a theory T_1 fails to accommodate certain facts (because no predictions about those facts can be derived from T_1) and another theory T_2 is then introduced to accommodate those facts, then, according to weak comparative predictivism, those facts would count as novel with regard to T_2 – so long as there are no other theories accommodating the facts. In strong predictivism, those same facts would count as novel with regard to T_2 only if T_1 (or some other theory) predicted the absence of those facts. Likewise, comparative predictivism is indifferent to the way in which a theory was constructed. So the aforementioned facts are novel with regard to T_2 even when those facts were used to construct T_2. Ladyman (1999) has pointed out that comparative predictivism seems to reward theories for being 'first' in accommodating certain evidence. However, whether or not a theory is the first to explain or predict a phenomenon may be due to highly arbitrary happenstance. For example, it might be due to higher publication rates of the journal in which the theory of interest appeared; other theories entailing the phenomenon of interest (or entailing its absence) might simply not make it into publication on time. But such contingencies should not influence our epistemic appraisal of a theory.[13]

---

[12] Mayo (1996, 208, fn. 19) reports that Popper, in personal communication, indeed committed himself to the strong version.

[13] Harker himself recognizes the problem but claims that he can avoid it by restricting his account to theory *preference* and talk about *progress* rather than truth (Harker 2008, 444–6). However, if such a view is supposed to ground realism, then preference and progress must be epistemic, which would make Harker's view subject to the same objections as Leplin's. Cf. Chapter 2, fn. 6 and 8.

It is also worth noting that comparative predictivism, in contrast to use-novelty in particular, is no longer informative of whether or not a hypothesis is ad hoc. With comparative predictivism, on the other hand, a theory can be novel – namely when it entails evidence which its competitors say nothing about or is prohibited by its competitors – and still ad hoc. That this is problematic can be seen with the example mentioned by Harker himself, namely Fresnel's explanation of straight-edge diffraction. As Worrall (1976, 193) points out, competitor theories were indeed able to account for such phenomena. They just did so in a very 'contrived' way. The same is true of the aforementioned alternative explanations of light-bending by the Newtonians (1996). Generally speaking, it always seems possible to save the phenomena in *some way*. At the very least, the comparative account of prediction needs to be amended by an account of ad hocness.

3.5   Mendeleev's Periodic Table and the Prediction of Chemical Elements

Having reviewed the most influential views of novel success, I now want to assess particularly the temporal and the use-novel account on the basis of one important example of novel success, namely Dimitri Mendeleev's successful prediction of various chemical elements on the basis of the periodic table he proposed in 1869.

The question of whether Mendeleev's successful predictions contributed to the acceptance of his periodic table of chemical elements is hotly debated (Maher 1988; Howson and Franklin 1991; Brush 1996; Scerri and Worrall 2001; Schindler 2008b). Two pieces of historical evidence are particularly contentious: the award of the Davy medal to Mendeleev and citation data. Let us briefly consider them in turn.

Maher (1988), Lipton (1991/2004), and Barnes (2008) have cited the award of the Davy medal to Mendeleev in 1882 by the Royal Society as evidence for temporal predictivism. By the time the medal was awarded, two chemical elements (gallium and scandium) had been discovered in a period of 11 years, which Mendeleev had predicted when first proposing his periodic table (1871). Maher, Lipton, and Barnes seem to treat the fact that the Davy medal was not awarded to Mendeleev *before* any predicted elements were discovered as evidence for predictions playing a vital role in the appraisal of Mendeleev's table. Yet this alleged support for temporal predictivism can quickly be dismissed. As Scerri and Worrall (2001) point out, there is no mention whatsoever in the Davy medal laudation of

Mendeleev's successful predictions of new chemical elements. Rather, emphasis is put on the 'marvelous regularity' in the properties in two series of already-known elements as revealed by Mendeleev's periodic table. Furthermore, the Davy medal was jointly awarded to Lothar Meyer for his table of chemical elements; contrary to Mendeleev, Meyer did *not* make predictions about any new elements.

The second piece of evidence temporal predictivists cite in support of their view is the fact that Mendeleev's periodic table received a real boost of citations in the period of 1876–1885, i.e., shortly *after* the first successful discovery of one of Mendeleev's predicted chemical elements in 1875 (Barnes 2008). In that period, the number of journal citations in fact tripled compared with the number of citations Mendeleev's periodic table received in the first four years after its publication (and before the confirmation of one of his predictions). Could there be any clearer evidence for temporal predictivism than that? But a closer look at the citation data sheds some doubt on this position (see Brush 1996). In the years in which gallium (1875), scandium (1879), and germanium (1886) were discovered, and in the respective subsequent years, the journal citations of Mendeleev's periodic table which also mentioned 'confirmation of Mendeleev's predictions about new elements' were 15 out of 27, 13 out of 21, and 7 out of 26, respectively (Brush 1996, 601). Obviously, a large number of those authors who discussed Mendeleev's periodic table – in fact more than half (53%) – did *not* refer to the successful predictions shortly after the respective elements were discovered. With Scerri and Worrall, one might think that 'it would be strange indeed for an author who had been drawn to Mendeleev's scheme in large part by its predictive success to make no mention of that success at all' (433). Thus, the bibliometric evidence for temporal predictivism in the case of Mendeleev's successful prediction of new chemical elements is at best equivocal. Brush also, more casually, has analysed contemporary chemistry textbooks and concluded that

> almost every discussion of the periodic law in nineteenth-century chemistry textbooks, including Mendeleev's, gives much more attention to the correlations of properties of the known elements with their atomic weights than to the prediction of new ones. (Brush 1996, 612; emphasis added)

In addition to the successful prediction of new elements, Mendeleev also made so-called contrapredictions. They will be the subject of the next section.

3.5.1 Contrapredictions

Successful contrapredictions (a term coined by the historian Stephen Brush) are predictions of theories that induce *corrections* to empirical results that were hitherto falsely accepted as correct. Interestingly, Mendeleev himself highlighted contrapredictions as of utmost importance in the assessment of the periodic table:

> Where, then, lies the secret of the special importance which has since been attached to the periodic law, and has raised it to the position of a generalisation which has already given to chemistry unexpected aid, and which promises to be far more fruitful in the future and to impress upon several branches of chemical research a peculiar and original stamp? ... *In the first place* we have the circumstance that, as soon as the law made its appearance, *it demanded a revision of many facts which were considered by chemists as fully established by existing experience.* (Mendeleev 1901, 475)

In Mendeleev's time, the atomic weight of an element was determined via an element's valence and 'equivalent weight'.[14] Accordingly, there were two sorts of contrapredictions Mendeleev made: (i) predictions that demanded changes to an element's presumed valency and (ii) predictions that demanded changes to an element's presumed equivalent weight. Mendeleev made several counterpredictions of types (i) and (ii), whereby most proposed atomic weight changes associated with (i) turned out to be correct and most proposed changes associated with (ii) incorrect (Smith 1976, 326ff.).[15] There were three principles on the basis of which Mendeleev made his contrapredictions (and in fact also his novel predictions): atomic weight ordering from lower to higher values, property resemblance of the elements in a particular group in the table, and 'single occupancy' of places in the table (Scerri and Worrall 2001). For Mendeleev, any discrepancy in the results that the application of each of those principles implied pointed to the re-assessment of atomic weight, property resemblance, or both. Interestingly, he never was prepared to violate the principle of atomic-weight ordering, despite genuine exceptions to it.[16] Whereas before mid-

---

[14] An element's equivalent weight is its relative weight as determined by the proportion with which it combines with hydrogen (and later oxygen).

[15] With regard to the 'placing' of the elements in the periodic table, however, the picture was reversed (cf. Smith 1976, 326ff.).

[16] These exceptions concern so-called pair reversals, i.e., cases in which two elements' places in a periodic table ordered by weight are reversed because of their chemical properties. For example, although iodine has a lower weight than tellurium, iodine should come after tellurium in order for both to be grouped with elements with similar properties. Mendeleev, trying to uphold the chemical-weight criterion, argued that the weights of the two elements should be adjusted accordingly. However, he did so erroneously: the pair reversal is genuine (tellurium weighs 127.6u and

1869 Mendeleev mostly re-positioned elements in the table while retaining the weight, after mid-1869, most of his re-positioning involved a change in atomic weight (326ff.). Let us briefly consider three examples of Mendeleev's successful contrapredictions.

Perhaps Mendeleev's boldest contraprediction concerns uranium. Uranium was first isolated in 1841 and was usually assumed to have an atomic weight of 120 and a valence of 2. But with this weight, one of the constraints of Mendeleev's periodic table (continuous decrease of valence from group IV to group VII) would have been violated, since uranium would then have to be placed in between the tetravalent tin (122) and trivalent antimony (118). Mendeleev proposed that the atomic weight for uranium be 240 by doubling its assumed valence, which turned out to be essentially correct (its accepted weight is 238u). In a later edition of the *Principles of Chemistry*, Mendeleev wrote that 'uranium ... has played a prominent role in the confirmation of the periodic law, because with the recognition of this law a change in its atomic weight was called for, and was proved valid'. For Mendeleev, this served '*as convincing evidence of the generality of the periodic law*' (both quotations from Smith 1976, 335; emphasis added).

Mendeleev also suggested a correction to beryllium, which, on the basis of a number of experiments, had been determined to be trivalent and to possess an atomic weight of about 13.5–14.0 (cf. Scerri 2007, 128). Again, according to Mendeleev (1901), this weight 'became generally adopted and seemed to be well established' (484). However, as he noted, 'there was no place in the system for an element like beryllium having an atomic weight of 13.5' (484). The chemical properties of beryllium were just too dissimilar from nitrogen (atomic weight of 14) to place it next to it. Instead, Mendeleev surmised that beryllium was bivalent because it could then be placed in between lithium (7) and boron (11) in group II. After 'a divergence of opinion [which] lasted for years' (484), Mendeleev was vindicated in 1889.

Ironically, the successful prediction of gallium, the first discovered element to confirm any of Mendeleev's novel predictions, is in fact a partial contraprediction. In 1875, apparently without being aware of Mendeleev's prediction, the French chemist Lecoq de Boisbaudran discovered gallium. He determined gallium's density as 4.7 g/cm^3. Mendeleev, however, had predicted a significantly higher value,

iodine ca. 126.9u, although the former's number is 52 and the latter's 53). See also Scerri (2007, 130–1).

namely about 6.0 g/cm^3. Only after becoming aware of that contra-prediction, Boisbaudran realized that his original measurements had been contaminated with metallic sodium (which he used as a reducing agent in the isolation of free gallium). And indeed, Boisbaudran now gained a density of 5.935 g/cm^3, very much in agreement with Mendeleev's contraprediction. Mendeleev was thus able to claim later that *without* the periodic table 'nothing would have pointed to the incorrectness of this [earlier] determination, nothing would have prompted the verification of the difficulty obtained and separated gallium' (1901, 262).

What kinds of beasts are contrapredictions? With Brush (1996), but contrary to Scerri (2007), I believe that we should treat contrapredictions as a particular form of temporally novel predictions. Just like in temporally novel predictions, and contrary to accommodations, in contrapredictions it is not known at the time a particular prediction is made whether the contraprediction will turn out to be correct. Contrapredictions and tempo-rally novel predictions differ of course in that in the former there is, and in the latter there is not, evidence relevant to the assessment of the prediction. But I don't think that this is a significant difference. This judgement is in fact supported by Mendeleev's own writings. With regard to the successful contraprediction of the weight of beryllium, Mendeleev noted that he took it '*as important in the history of the periodic law as* the discovery of scandium' (e.g., 1901, 485). Indeed Mendeleev, when commenting on the reasons for the widespread acceptance of the periodic table, highlighted contrapredictions *and* temporal predictions (Mendeleev 1879, 231). I do not think that we have any reason to suspect that Mendeleev was somehow deluded about the reception of his table in the scientific community. And with regard to the aforementioned citation data, it should be pointed out that they do not include data on contrapredictions. Brush's aforementioned observation that chemist textbooks did not emphasize novel success also concerns only newly discovered elements, not contrapredictions. If contrapredictions are counted as temporally novel predictions, the quoted citations should thus be treated as lower bounds. Before I reveal how I think we can make sense of all of this, let us discuss how (if at all) some of the accounts we have covered in this chapter can deal with this case.

3.5.2 *Mendeleev's Periodic Table and Accounts of Novel Success*

According to Worrall's account of use-novelty, evidence E that was used (or needed to be used) in the construction of a theory T that entails E lends little or no support to T. We can immediately see that this view would have a radical implication for the confirmation of Mendeleev's periodic table: the table would receive little or no confirmation from the 55-plus elements on the basis of which Mendeleev erected the periodic table (Barnes 2005). But that is absurd. The periodic table derives much of its force from the fact that it relates *all* elements to their atomic weight (and, later, atomic number). And without the elements which Mendeleev used to construct his table, there would have been no periodic table in the first place.

Prima facie, Hitchcock and Sober's local-symptomatic predictivism appears to nicely capture Mendeleev's successful contrapredictions. That is, in view of the wrong weights of some of the elements prior to Mendeleev's corrections, one might say that there was a risk for anyone constructing a periodic table that they would seek to accommodate the wrong chemical weights, i.e., that they would overfit the data. Only after Mendeleev successfully predicted new data (regarding both new and known elements) did it become clear that Mendeleev's periodic table did not overfit the data. However, beyond this, Hitchcock and Sober's account runs into a number of difficulties.

As noted earlier, Hitchcock and Sober's conclusions apply only to models/theories which possess empirical parameters and which can be fixed on the basis of data. What are the empirical parameters of Mendeleev's table supposed to be? The only plausible candidates appear to be the slots for the chemical elements with a certain chemical weight in the periodic table. If construed in this way, however, the *number* of parameters seems to make little difference to whether or not Mendeleev's table overfitted the data: whether the periodic table reserved more or less slots for the chemical elements (known or not known), this seems to have very little relevance to whether or not the predictive success of Mendeleev's table was likely. But again, this is what Hitchcock and Sober have argued for: the more free parameters there are in the model, the more likely the model is to overfit the data.

I myself have defended a local-symptomatic predictivism with regard to the status of the Mendeleev's successful predictions (Schindler 2014a). In contrast to Hitchcock and Sober, however, I do not assert any *intrinsic* connection between predictive success and some theoretical virtue, such as simplicity (in their account, measured in terms of the number of free parameters). Similarly, I do not claim that predictive success is a *more likely* indicator of the theory in question having successfully identified a systematic pattern, or of a theory's unifiedness, as suggested by Lange (see Section 3.3.1; cf. Harker 2006, 436). But I do think that in some contexts predictive success can be indicative or symptomatic of another theoretical property. With regard to Mendeleev's periodic table, I have argued more specifically that Mendeleev's predictive success was indicative of the truth of the *explanatory coherence* of the periodic table, which amounted to the principle that the properties of *all* chemical elements ought to be explained in a minimal, yet manifest, sense by their place in the periodic table, which in turn was determined (in large part) by their chemical weight (and later by their atomic number) (cf. Schindler 2014a). By virtue of its explanatory coherence, Mendeleev's periodic table thus gave *good reasons* for believing that a particular element in the table *ought to* possess a particular weight and particular properties. This seems to accord with Mendeleev's own thoughts on the matter:

> Before the promulgation of this law the chemical elements were mere *fragmentary*, incidental facts in Nature; *there was no special reason to expect the discovery of new elements*, and the new ones which were discovered from time to time appeared to be possessed of quite novel properties [i.e., without apparent relation to properties of other elements]. (Mendeleev 1889, 648; emphasis added)

It was Mendeleev's great achievement that he put the explanatory coherence of the table above apparent empirical contradictions: previous attempts to classify the elements had never resulted in one coherent periodic table but only in patchworks of several periodicities that remained unrelated (cf. Smith 1976; Scerri 2007). Mendeleev himself compared his approach to the one of his predecessors: 'Formerly [the periodic law] was only a grouping, a scheme, *a subordination to a given fact; while the periodic law furnishes the facts*' (Mendeleev 1879, 231; emphasis added). Yet before Mendeleev's predictions were confirmed, chemists could of course not know whether the coherence identified by Mendeleev was real or not; they could not know this on the basis of the already-known facts, as the principle went beyond those. That is why the novel success, in this context, was of special importance: it demonstrated that the explanatory coherence of the periodic table identified by Mendeleev was real.

3.6   Conclusion: Novel Success and Nagging Popperian Intuitions

Popular accounts of novelty such as Worrall's and Hitchcock and Sober's relate novel success to parameter freedom. In Worrall's account, a piece of evidence E is novel with regard to a theory T when there is no free parameter in T which needs to be fixed to accommodate E. But then, as we saw, Mendeleev's periodic table would have gained very little confirmation from the chemical elements on the basis of which he erected his periodic table. There is also a question as to why parameter freedom is such a bad thing. On Worrall's account, it seems, parameter freedom is bad because the theory is guaranteed to accommodate the evidence in question. This in turn is bad because it decreases the risk of the theory being shown to be false. Worrall's account – as he himself admitted at some point – is thus driven by a genuinely Popperian intuition, which I shall comment on later. On Hitchcock and Sober's account, parameter freedom is bad because it reduces the predictive success of a data model. And yet it is an entirely open question how this rationale could be extrapolated to higher-level theories which are two steps removed from the data.[17]

Like Hitchcock and Sober, I do not believe that novel success is always better than the accommodation of non-novel facts (in the temporal or heuristic sense). Accordingly, I do not think that realism should be based on such a view. Unlike Hitchcock and Sober, I do not think that we can argue for some kind of intrinsic connection between some properties of a theory (such as simplicity) and novel success. Unlike Musgrave and (to some extent) Harker, I do not believe that comparative predictivism is a viable view (for reasons mentioned earlier). I therefore advocate a shift away from philosophers' obsession with novel success towards concerns that underlie some predictivist theses (in particular Worrall's weak version of use-novelty) and, specifically, concerns regarding ad hocness. I will myself provide a systematic analysis of ad hocness in Chapter 5. The upshot of my analysis will be that judgements that a hypothesis is ad hoc have something to do with a lack of explanatory coherence, something that Mendeleev's periodic table did possess. Because contemporary realism is heavily based on predictivism – as explained in the introduction to this chapter – my proposed shift in focus will also have to result in a shift in realist commitments.

Before we proceed, however, let me briefly comment on an intuition which hasn't figured so explicitly in debates about novel success but which may nevertheless explain the appeal of novel success: the 'Popperian' intuition that with temporally novel predictions, but not with

---

[17] Hitchcock and Sober's account might make sense for climate modelling. Frisch (2015) argues that climate models are epistemically opaque in that it is often not clear which features of the model are responsible for the empirical success of the model. Novel success, according to Frisch, can boost our confidence that the model has actually correctly identified features of 'mechanism' underlying climate change.

accommodative success, theories 'stick their necks out' and 'take a risk' of being refuted. Given the greater risk, theories should accordingly be awarded with a greater degree of confirmation.[18]

Let us contemplate what the Popperian intuition would mean when fully spelled out. For the intuition to have any traction, and for greater risk to result in a greater degree of confirmation, we need to presuppose more of the Popperian picture. That is, we need to – somewhat paradoxically – presuppose that science seeks to maximize our chances of being wrong and that this is something good. Only then would it make sense for us to confer more credit to a theory that takes a greater risk of being false. However, I (and many others) think the Popperian picture is thoroughly mistaken. Science is not about increasing our chances of getting things wrong, but rather about increasing our chances of getting things right. Moreover, I believe it is about getting things right in a non–ad hoc fashion. Again, a focused concern with what it means for a hypothesis to be ad hoc is in order. But first let us turn to another form of theoretical fertility.

---

[18] Strictly speaking, of course, this intuition is only partially Popperian, as Popper did not believe in confirmation.

CHAPTER 4

Theoretical Fertility without Novel Success

In Chapter 3 we discussed novel success as the most popular form of a theory's fertility. As I pointed out, novel success has played a crucial role not only in theory appraisal, but also in the realism debate: lately, the realist has been willing to commit to the truth of only those theories which have been able to generate novel success. The general rationale is that other kinds of empirical success are too cheap: any piece of evidence can be accommodated in an ad hoc fashion.

Interestingly, it has been argued in a string of articles by McMullin (1968, 1976, 1983, 1984, 1985) that a theory can be fertile even if it does not generate novel success and, moreover, even when it is modified in a way that would count as ad hoc under the standard notions of novel success (namely, temporally novel and use-novel success).[1] Although standard realists would not commit to such theories being true, McMullin has claimed that what I shall refer to as M-fertility, provides 'perhaps the strongest grounds for the thesis of scientific realism' (McMullin 1985, 264).

In this chapter I review and assess in detail M-fertility and its realist rationale. I argue that although immune to objections that have been raised against M-fertility being a theoretical virtue in the first place, M-fertility per se does not support realism. I do so by closely re-examining McMullin's preferred example of the development of the Bohr–Sommerfeld model of the atom. At the same time, I suggest, this example can still teach us important methodological lessons about theory appraisal.

I proceed as follows. Section 4.1 introduces M-fertility and reviews the objection, raised by Nolan (1999), that M-fertility is not a virtue. I reject Nolan's claim that M-fertility is a virtue only insofar as it reduces to novel success, amongst other things, by emphasizing an aspect of M-fertility that

---

[1] Realists usually appeal to the temporal or use-novel success notion. They generally do not appeal to the parameter-freedom notion.

is entirely neglected by Nolan, namely the idea of de-idealization (4.1.2). In Section 4.1.3, I point out that M-fertility, and in particular the idea of de-idealization, can be used to make sense of Lakatos's notoriously vague idea of a research programme's positive heuristic. In Section 4.2, I argue in detail that the development of the Bohr–Sommerfeld model of the atom – McMullin's and Lakatos's preferred example – does not support McMullin's idea of de-idealization. In Section 4.3, I contrast the fertility of the Bohr model with the lack thereof in the case of the kinetic theory of heat and the so-called specific heat anomaly. In Section 4.4, I conclude my discussion.

4.1 McMullinian Fertility and Nolan's Challenge

In one of his many papers on the topic, McMullin characterizes a theory's fertility as the ability to 'cope with the unexpected' (McMullin 1976, 423), in particular 'as new evidence becomes available' (1968, 391). In another place, McMullin specifies that

> the theory proves to have the *imaginative resources . . . to enable anomalies to be overcome and new and powerful extensions to be made.* Here it is the long-term proven ability of the theory or research program to *generate fruitful additions and modifications* that has to be taken into account. (McMullin 1983, 16)[2]

To put it schematically, at first pass, a theory T is M-fertile if T has resources for accommodating evidence E, inconsistent with T, by suggesting modifications of T so that a modified version of T, namely T*, can accommodate E. This definition will be made more precise in what follows.

It is easy to see that M-fertility is incompatible with both of the standard notions of novel success. For temporally novel success, evidence E must minimally be unknown at the time at which theory T is devised. E must also be anticipated by T. But that is clearly not the case for what McMullin has in mind. In M-fertility, modified T* accommodates evidence that could not be accommodated by the original T (an anomaly for T is by definition not anticipated by T). Use-novelty is also violated in M-fertility: T is modified *in order to* accommodate E. Hence, insofar as M-fertility is a virtue, it is a virtue that is independent of the virtue of novel success, at

---

[2] McMullin also compares the function of models to that of metaphors, speaks of the model's imaginative resources as 'metaphorical power', and refers to the changes of the model in response to anomalies as 'metaphorical extension' (cf. McMullin 1984, 30f.).

least when it comes to temporal success and weak use-novel success. What about strong use-novel success, or other notions that construe novel success in terms of parameter-freedom (Chapter 3)? It is not immediately clear how 'fruitful additions and modifications' in response to anomalies could figure in them. Obviously, since in M-fertility original T cannot accommodate E, T will not possess any free parameter for E. So, under Worrall's strong notion of use-novel success, E would not count as ad hoc. On the other hand, since T cannot accommodate E, E cannot count as use-novel success either. So even though M-fertility may not be inconsistent with strong use-novel success, it does not seem to be captured by it. Thus, by the lights of standard notions of novel success, M-fertility is not a virtue.

Before we proceed, let us also note that McMullin distinguishes between P(roven) and U(nproven) fertility. It is the former with which McMullin is mainly concerned. U-fertility McMullin describes as a theory's 'heuristic potential', or the 'as-yet unexplored heuristic possibilities of the theory' being high (McMullin 1976, 424). That is, a theory that is U-fertile has not yet been shown itself to be capable of accommodating anomalies but might nevertheless hold the potential of doing so in the future. For McMullin, U-fertility is a non-epistemic virtue and therefore of little interest to him.[3] We shall follow McMullin and lay our focus on P-fertility. Accordingly, unless otherwise indicated, we will read M-fertility as P-fertility by default.

4.1.1   Nolan's Challenge

Nolan (1999) has questioned that there is a feasible rationale for viewing M-fertility as a virtue 'in its own right'. Nolan writes:

> On the face of it, it can be a little hard to see why the liability of a theory to require improvements, or to raise new problems, should be considered a good thing. It is as if 'Faces many problems' or 'Could do better', or 'Much room for improvements' are high praise on the report card of a theory. Surely being in difficulties is not what we would expect of an ideal theory? (267)

'Kuhnian reasons' for thinking that M-fertility might be a virtue – such as 'opportunity to do science', intellectual challenge (267), and 'employment prospects' (268) – Nolan rejects as unconvincing, and rightly so.[4] Although

---

[3] McMullin uses the idea of U-fertility mainly to describe and criticize Lakatos's account. See Section 4.1.3 for a comparison of the two accounts.

[4] Denoting such sociological aspects of science as 'Kuhnian' seems slightly unfair. Contrary to common misinterpretations, Kuhn thought that there was much more (cognitively) involved in

these reasons might be good pragmatic reasons for valuing a theory, they are not epistemic ones. And it is those we ought to be after in the philosophy of science.

Nolan does not say that M-fertility has no value. However, he denies M-fertility a value 'over and above the virtues which in fact its value rests [upon]' (265). In other words, Nolan is a reductionist about M-fertility. Although Nolan, like McMullin, makes P-fertility the main focus of his discussion, his reductionism extends also to U-fertility. For Nolan, U-fertility is 'the potential to *advance*, or *improve*, or *progress*' and is valuable 'as a means, rather than an end', where the end is some other virtue (271; original emphasis). Nolan is fairly unspecific about the virtues U-fertility might be a means for. However, he does mention in passing 'increasing strength, increasing predictive power and accuracy, increasing unification' (270). U-fertility is, then, the 'chance that a close successor or a later theory-stage will be significantly better than the current one' with regard to these (or other) virtues (271). With regard to P- or M-fertility (recall that we treat those as equivalents), Nolan is much more specific: he attempts a reduction of M-fertility to novel success. Given that, as mentioned earlier, M-fertility and novel success are incompatible, this should come as a surprise. Nolan's reduction therefore requires a little tweak.

Nolan invites us to consider a sequence of related theories V_1, V_2, V_3, etc. Each of those gets revised after the fact, we suppose, on the basis of some anomalies, so that they each eventually accommodate those anomalies (273). Crucially, Nolan proposes that we associate with these specific theory developments a meta-hypothesis V, for example the meta-hypothesis that 'some atomic theory of gases is true' or that 'the best available theory is going to lie in the tradition of V'. Now,

> the [meta-]hypothesis that the V-series of theories is on the right track (or some V-like theory is true, or close to the truth, or the best theory of the phenomena available) may be doing very well nonetheless, and *may even be confirmed by the evidence which refutes specific versions*. (276; emphasis added)

In other words, evidence that looks anomalous from the perspective of the specific V-theories can be viewed as successes from the perspective of the meta-hypothesis V. What is more, since the meta-hypothesis came to life with the first specific V theory (at the latest), we can view all those anomalies for the specific V-theories as novel evidence for the meta

---

the pursuit of normal science than just opportunism. There is however some textual evidence that Kuhn, at least in his later work, thought of fertility *also* in sociological terms (Kuhn 1977a, 322, fn. 6).

hypothesis V: 'the fact that V-like theories got better and better counts as novel confirmation for the theory that V-like theories are good' and 'on the right track' (276).[5] Or so Nolan suggests.

Several things about Nolan's proposal should strike one as dubious. First of all, the sort of novel success Nolan claims for meta-hypotheses is a pretty thin one. Normally, we have reason to be impressed with a theory's novel success because the predicted phenomena are *precisely* as predicted by the theory. But Nolan's meta-hypotheses give us no such reasons. On the contrary, Nolan's meta-hypotheses are extremely vague: they are compatible with all sorts of evidence. Relatedly, although it is true that 'the fact that this general claim [i.e., the meta-hypothesis] was advanced *before* the history of specific versions renders it able to have novel confirmation' (278), this is not sufficient for novel success. That is, it is not sufficient for E to count as novel success for T that E comes chronologically before T; E must be *predicted* by T. But that is clearly not the case. The meta-hypothesis does not predict any of the anomalies faced by the specific versions. If it did, then the anomalies would not count as anomalies in the first place! So even if one were to be persuaded that the anomalies of the specific V-theories might be evidence for the meta-hypothesis V, one would be at a loss as to why these pieces of evidence ought to count as novel success for the meta-hypothesis. Second, Nolan's appeal to meta-hypotheses is highly arbitrary. Nolan offers us no constraints for choosing 'some atomic theory of gases is true' rather than meta-hypotheses at even higher levels of abstraction, as for example 'all matter is made up of atoms' or 'unobservables exist'. In comparison to *those* hypotheses, *any* accommodative success in the realm of chemistry and physics concerning atoms and molecules would count as temporally novel success on Nolan's account. But that is clearly absurd. Thirdly, Nolan pretty much ignores the fact that McMullin *does* offer a rationale for why M-fertility ought to be a virtue.[6] Let us have a closer look.

4.1.2 Non–Ad Hoc Modifications through De-idealizations

For McMullin, not any old modification of a theory may count in its favour, when it comes to the gauging its empirical success. Minimally, modifications of a theory in response to an anomaly are only allowed when

---

[5] Nolan suggests that the meta-hypotheses may be construed as some kind of 'general theory' instead (278f.). This difference is minor and shall not concern us here.

[6] See Segall (2008) for another criticism of Nolan's proposal.

they are not ad hoc (e.g., McMullin 1985, 264). And modifications of a theory in response to anomalies are not ad hoc, in turn, when the modifications result from *de-idealizations* of the original theory. In a good theory, McMullin surmises, these de-idealizations are suggested by the original theory itself:

> If the model is a good one, these processes [of self-correction] are not ad hoc; they are suggested by the model itself. Where the processes are of an ad hoc sort, the implication is that the model is not a good one ... (McMullin 1985, 264)

The idea of de-idealization is therefore to be understood thus: although scientists are very well aware of the fact that the assumptions underlying an initial model/theory might be unrealistic, they leave out complications in the first steps for the sake of simplicity (cf. McMullin 1968, 394). In sum, then, M-fertility may be defined as a theory's ability to accommodate anomalies in a non–ad hoc fashion by means of *de-idealization* of its original simplifying assumptions.[7] Put schematically, a theory T is M-fertile if an anomaly E of T can be accommodated by de-idealizing T to T* so that T* entails (or makes likely) E. Why is this a virtue? There are two, fairly independent, reasons to be found in McMullin's work.

First, M-fertility is a virtue because it allows the accommodation of apparent anomalies in a non–ad hoc fashion, and non–ad hocness clearly is a methodological desideratum. There is a *separate* question of whether non–ad hocness increases the chances of a theory being true. This may be controversial (but see Chapter 5). What is uncontroversial, however, is that a theory that is ad hoc with regard to some evidence should receive less confirmation from that evidence than a theory that accommodates this evidence in a non–ad hoc way. Since confirmation clearly is an epistemic matter, non–ad hocness should be too (see Chapter 5). Second, M-fertility is a virtue on McMullin's account, because it indicates to us that the original model we started out with got something fundamentally right about nature:

---

[7] McMullin, on several occasions, attributes a theory's fertility to the model associated with a theory. McMullin has a peculiar view of models. For him, '[t]he theory is derived from the model ... not the reverse ... [T]he theory is about [a particular] model and about nothing else' (McMullin 1968, 389). At least the latter part of this view seems to come close to Cartwright (1983)'s 'prepared descriptions' of real systems, which she believes theories are about, rather than about the real system itself. At the same time, McMullin (389) is quite adamant in his rejection of some of Cartwright's (antirealist) conclusions. The differences between models and theories shall not concern us here.

This technique [of de-idealization] will work only if the original model idealizes the real structure of the object. (McMullin 1985, 261)

Without such a [rough] fit [between the original model and the real structure of the object], there would be no reason for the model to exhibit this sort of fertility. This gives perhaps the strongest grounds for the thesis of scientific realism. (264)

In other words, the approximate truth of the original model explains why the changes suggested by the model itself help it overcome anomalies. Although this resembles the standard explanationist defence of realism, McMullin's explanandum is different (M-fertility instead of novel success). Thus, there is not just one but actually two rationales McMullin offers for M-fertility. It's quite surprising that Nolan, in his discussion of M-fertility, makes no mention of either of these.

4.1.3   *M-Fertility and Lakatos's Positive Heuristic*

Before we proceed with the assessment of the rationales of M-fertility offered by McMullin, it is worth dwelling on their strengths: not only do they make it plausible that M-fertility is a virtue in its own right, but, as we shall see in the current section, they also seem to offer clarification with regard to related accounts, such as Imre Lakatos's ideas of research programmes. The comparison will also give us a better sense of the role of anomalies in McMullin's account.

For Lakatos, research programmes (basically a theory development over time) consist of a 'hard core' of 'irrefutable' assumptions and a 'protective belt' of auxiliary assumptions which 'bear the brunt of tests and get adjusted and re-adjusted, or even completely replaced, to defend the thus-hardened core' (133). The so-called 'negative heuristic' of research programmes 'forbids us to direct the *modus tollens* at this "hard core"', in contrast to the 'positive heuristic' that guides scientists in the construction and modification of the protective belt. Lakatos describes the positive heuristic also as providing 'instructions' for building 'ever more complicated models simulating reality' (133–5). Although this is very vague, Lakatos's examples suggest that he has in mind something very similar to McMullin's idea of de-idealization. Newton, for example, first assigned infinite mass to the sun as the central body in the solar system in order to calculate planetary orbits on the basis of the law of gravitation. Later, he relaxed this assumption so that the sun and the planets would receive a common centre of gravity. Furthermore, Newton first treated the planets and the sun as mass-points and only later as mass-balls. All these changes were reasonable and expected changes – just like they are in the de-idealizations McMullin talks about. Indeed, McMullin seems to have the same thing in mind as Lakatos. A difference between McMullin's and Lakatos's accounts, however, is the emphasis that Lakatos places on novel success in the evaluation of research programmes and how tolerant he appears to be of anomalies.

Whereas in M-fertility anomalies 'trigger' de-idealizations, anomalies have no place in the development of a research programme in Lakatos's view. Ideally, for him, '. . . the positive heuristic forges ahead with almost complete disregard of "refutations" . . .' (137). Indeed, Lakatos goes so far as saying that 'relatively few experiments are really important' for the development of a research programme, as long as it is progressive (151). In contrast, when scientists do start caring about modifying the research programme so as to accommodate anomalies, it is degenerating:

> Which problems scientists working in powerful research programmes ration-ally choose, is determined by the positive heuristic of the programme *rather than* by psychologically worrying (or technologically urgent) anomalies. The anomalies are listed but shoved aside in the hope that they will turn, in due course, into corroborations of the programme. *Only those* scientists have to rivet their attention on anomalies who are either engaged in trial-and-error exercises or who work in a degenerating phase of a research programme when the positive heuristic ran out of steam. (Lakatos 1970a, 137; emphasis added)

Lakatos also requires *consistent* theoretical progress for a progressive research programme; that is, he requires that any modification of a research programme result in novel predictions (134). And although Lakatos demands only *intermittent* empirical progress – i.e., occasional confirmations of the novel predictions – ultimately, any research pro-gramme must produce novel success at some point for it to be considered progressive (134). Once a research programme does manage to produce novel success, this will have a positive, retrospective bearing on the assess-ment of the *entire* research programme, regardless of how many anomalies the programme has encountered hitherto: 'a long series of "refutations"' can therefore be turned 'into a resounding success story' (134).

Lakatos is quite explicit that the notion of novelty he has in mind is temporal novelty. That is, the same piece of evidence can count either towards the progressiveness or the degeneracy of a research programme, depending on whether that piece of evidence was predicted (in the tem-poral sense) by the research programme or whether it was already in place when the research programme was developed (151–2). Lakatos's demand that theoretical progress be 'consistent' is in fact motivated by the value he places on novel success: only a research programme that 'forges ahead', before the evidence relevant to its predictions are gathered, can give itself a chance of having *temporally novel* success.

Thus, although there are clear similarities between M-fertility and Lakatos's progressive research programmes (in particular, theoretical change through de-idealization) a crucial dissimilarity is Lakatos's insistence on (ultimate) novel success as a necessary condition for theoretical progress. On McMullin's account, by contrast, since changes to the theory follow the encounter of anomalies, both temporally novel success and use-novel success criteria are violated. Of course, if de-idealization is what justifies the changes to a theory in both McMullin's and Lakatos's accounts, as we have argued, then it seems merely a contingent matter that in some cases changes to the theory are made after the encounter of anomalies and in other before anomalies are faced. At bottom, one might say, there is simply some kind of intrinsic necessity to de-idealize, regardless of whether the evidence that is being accommodated comes in before or after the de-idealization is actually made. Lakatos's insistence on temporally novel success seems misplaced. Indeed, if the rationale for M-fertility is sustainable, then the realists' obsession with novel success seems likewise gratuitous.

4.2 The Bohr–Sommerfeld Model of the Atom

Both McMullin and Lakatos have used the Bohr–Sommerfeld model of the atom as a paradigmatic example of their accounts. This model was first expounded by Niels Bohr in his famous trilogy in the *Philosophical Magazine* in 1913 and substantially developed by Arnold Sommerfeld in the period of 1913–1925 (McMullin 1968, 393–5; Lakatos 1970a, 146ff.; McMullin 1985, 259ff.). In the remainder of this chapter I want to argue that this example, despite what McMullin and Lakatos think, does not support the idea of M-fertility. In particular, the historical development of the model is not captured by the idea of de-idealization, which is central to M-fertility.

Conveniently, McMullin and Lakatos divide the historical development of the Bohr–Sommerfeld model into similar de-idealization stages, each of which allowed the model to accommodate certain anomalies, and each of which, according to Lakatos (1970a), 'was planned right at the start' and which, according to McMullin, *had* to happen (1968, 394–5):[8]

---

[8] The similarities of McMullin's and Lakatos's accounts are indeed striking. Who was first and who might have been inspired by the other is not the subject of this chapter. For an attempt to work out

1. *Assigning finite mass to an initially infinite mass atomic nucleus.* This change helped account for the Pickering–Fowler spectral lines of ionized helium.
2. *Turning circular electron orbits into elliptic orbits.* This allowed the model to account for the so-called Stark effect, i.e., the splitting of spectral lines in an electric field.
3. *Adding relativistic effects to the elliptic electron orbits.* This accounted for the fine structure of hydrogen.

A final stage in the development of the Bohr–Sommerfeld model is mentioned by both McMullin and Lakatos but treated slightly differently from the previous three stages (details to be discussed further on).

4. *Introducing the concept of electron spin.* This was as suggested by Goudsmit and Uhlenbeck in 1925, in response to the anomalous Zeeman effect.

As we shall see in the following discussion, none of these stages supports the idea of accommodation of anomalies/novel predictions through de-idealization. Thus, contra McMullin, none of these stages can support realism, as per McMullin's arguments.

4.2.1   Stage 1: Finite Mass Nucleus and the Pickering–Fowler Series

4.2.1.1   The Facts

In 1897, the American astronomer E. Pickering measured several spectral lines emitted from the star ξ-Puppis. Because those lines converged to the same limit as the Balmer series, which described the spectral lines of hydrogen, they too were ascribed to hydrogen. In 1912, the British astrophysicist A. Fowler managed to reproduce these lines in the laboratory, with a discharge tube containing a mixture of hydrogen and helium. Contrary to the Bohr model, these additional lines seemed to suggest half-integral quantum numbers, whereas the Bohr model only allowed for integral ones. Bohr, however, speculated that Pickering's and Fowler's measurements were not caused by hydrogen but rather by ionized helium. He therefore calculated the expected lines with an atomic model containing two protons and one electron, and was able to derive Pickering's and Fowler's results. Indeed, Bohr predicted that the relevant spectral lines

---

differences beyond the ones emphasized here, see McMullin (1976). With regard to the Bohr model, McMullin's and Lakatos's accounts are not equally detailed in all respects. I shall use something like the best reconstruction.

would be measured even when chlorine instead of hydrogen was to be used as a catalysator. This prediction was confirmed by Evans on 4 September 1913. The remaining small discrepancy between Bohr's predictions for ionized helium and Fowler's data was largely removed by assigning finite mass to the atomic nucleus, allowing for a small motion of the nucleus around a common centre of mass due to the (minimal) attraction of the nucleus by the electron.[9]

4.2.1.2   Assessment

Lakatos calls Bohr's increase of the atomic number of the model a 'monster-adjustment', which he describes as 'turning a counterexample, in the light of some new theory, into an example' (Lakatos 1970a, 149, fn. 1). Lakatos tries to justify this 'monster-adjustment' retrospectively, by the ensuing novel success. More specifically, Lakatos singles out Bohr's prediction that the Fowler line should be observable in tubes containing no hydrogen at all, but only helium and another catalysator (Lakatos 1970a, 149, fn. 1). And indeed, it was reported that Einstein called this confirmation 'an enormous achievement' and concluded that 'the theory of Bohr must then be right' (von Hevesy, in a letter to Bohr on 23 September 1913, in Hoyer 1981, 531).

First of all, one may question whether increasing the atomic number of the model really constitutes a change of the model in the first place (let alone a 'monster-adjustment'). After all, the basic assumptions seem to be fairly unscathed; it is simply applied to a different chemical element, namely helium. Second, even if we were to accept this as a change of the model, it is hard to see how applying the model to a different element could count as a de-idealization. And third, it is at least questionable whether the successful prediction of the Fowler line in tubes containing no hydrogen can count at all as genuine novel success: both hydrogen and helium were of course known chemical elements at the time. Since the model was adjusted in order to accommodate the anomaly, we are here not even dealing with a use-novel prediction. In spite all of this, Bohr's theory was received extremely positively in some quarters, even before the discovery of the Fowler line. For example, Sommerfeld wrote to Bohr on 4 September 1913:

---

[9] In fact, following Fowler's reaction to Bohr's suggestion, Bohr proposed a further correction to his model, namely the velocity dependence of the mass of the electron (Bohr's letter to Fowler on 15 April 1914, in Hoyer 1981, 504ff.). This, to my knowledge, is the first occasion at which Bohr considered relativity corrections to his model in writing. See also the next stage discussed in the main text of this chapter.

> Although I am for the present still a bit skeptical about atomic models, your calculation of the constant [in the long-known Balmer formula] is nevertheless a great achievement. (Hoyer 1981, 603)

Overall, the idea of de-idealization (let alone some notion of novel success) appears not to underlie physicists' positive reception of the Bohr model with regard to the anomaly of the Fowler line.

4.2.2   Stages 2 and 3: Ellipses, the Stark Effect, and the Fine Structure

Both McMullin and Lakatos present the move from circular to elliptical orbits, largely associated with Sommerfeld's development of the model, as a matter of course – that is, as an almost necessary change of the model that was bound to happen even regardless of any anomalies. McMullin, for instance, writes that 'there is no reason to restrict the electron to circular orbits, since elliptical orbits are the normal paths for bodies under central forces of this type' (McMullin 1968, 394) and that

> [i]f one has a two-body system of the sort postulated, one will *have* to have nuclear motion, elliptical orbits, and relativistic mass effects, unless the laws of physics, known from elsewhere are assumed not to apply to the entities in the model. (394–5; original emphasis)[10]

Ironically, of course, the Bohr–Sommerfeld model *was* a model of entities which *did* behave very differently from what the laws of physics appeared to dictate elsewhere; stationary electron orbits *were* irreconcilable with the classical laws of physics. It is thus at least not unproblematic to appeal to the regular laws of physics to motivate changes to the model. One might even question whether the move from circular to elliptical orbits is accurately described as a de-idealization. After all, circular orbits are just a special case of elliptical orbits. One may perhaps say that circular motion is more unlikely than elliptical orbits insofar as in circular motion centripetal forces must precisely counterbalance centripetal acceleration. Circular motion would then be an idealization of elliptic motion in that, normally, conditions are such that centripetal forces and centripetal acceleration would *not* counterbalance each other in nature. At any rate, we shall see in a moment that elliptic orbits by no means replaced circular orbits; they complemented them. What is more, circular orbits were explicitly deemed *more likely* than elliptic orbits. None of this makes sense in a de-idealization story.

---

[10] Lakatos comments on this change that it proceeded 'as planned' from the outset of the Bohrian research programme (1970, 148).

What is also striking is that one finds no comments by Sommerfeld, Bohr, and others, which would give any indication that their motivation for introducing changes to the model might have something to do with de-idealization. Instead, as we shall see next, the *only* motivation for the introduction of ellipses by Bohr, who toyed with the idea in 1915, and by Sommerfeld, who delivered a detailed treatment in 1915–1916, was the accommodation of the fine structure of hydrogen and other elements. In other words, all that Bohr and Sommerfeld seemed to care about when designing their models was to accommodate the phenomena.

4.2.2.1 The Facts

Already in 1887 it was discovered by Michelson and Morley that one line in the spectrum of hydrogen was not a singlet, as described by the Balmer series, but rather a doublet. The importance of this finding remained unappreciated and appears to have received renewed attention only with the arrival of Bohr's model. In 1914, W. E. Curtis refined Michelson and Morley's observations and concluded that 'Balmer's formula has been found to be inexact' (Curtis 1914, 620) and that Bohr's derived spectral lines formula, corrected for the finite nucleus mass and the velocity dependence of the mass of the electron (cf. fn. 9), could not accommodate the results either (616). The suggestion was made that these results could be accounted for by the magnetic properties of the atomic nucleus. Bohr, in a discussion note, dismissed this proposal. Instead, he pointed to a mistake by Curtis and re-iterated the idea of relativity corrections for *circular* electron orbits. He admitted, though, that this would account only for a third of the deviations from the Balmer series, as observed by Curtis. Bohr conceded to Curtis that his earlier idea of explaining the doublets of hydrogen as 'not true doublets' and as resulting from the presence of an electric field was incorrect. He continued that 'there is perhaps another way of explaining the observed doubling without introducing new assumptions as to a complicated internal structure of the hydrogen nucleus'. What he was pondering was the supposition 'that we would obtain a doubling of the lines [of hydrogen] if the orbits are not circular'. In view of 'great number of new assumptions involved in such a calculation', he preferred to 'await more accurate measurements' (Bohr 1915).

Sommerfeld took a more proactive approach. In the same year in which Bohr wrote his discussion note, Sommerfeld presented a detailed treatment of elliptic orbits for the Bohr model to the Bavarian Academy of Sciences, which was then published the following year in the society's protocols (1916b) and in more extended form in Annalen der Physik (1916a). In the latter, Sommerfeld set out to 'extend significantly' Bohr's theory with a 'consideration of non-circular orbits' in order to 'shed light on the 'Sonderstellung' [i.e., peculiar status] of spectral lines of hydrogen', which consisted in hydrogen having just a single Balmer's series, whereas other elements had different line series (principal, subordinate, and combinatorial) and different line types (singlets, doublets, and triplets). His basic idea was this:

> According to the view to be presented here this is to be explained by the fact that [in the hydrogen spectrum] a number of [different] series coincide in the Balmer series [of hydrogen], i.e. each of [hydrogen's] lines can be produced in a number of different ways, *not only through circular motion but also through elliptic orbits* with particular eccentricities, as can be shown experimentally. (Sommerfeld 1916a, 4; emphasis added)

Interestingly, Curtis, although not mentioned by Sommerfeld, had developed a very similar idea, not on theoretical, but on comparative, grounds. The relevant passage is worth quoting in full for clarification:

> As [the Sharp series of hydrogen] have never been seen, however, it is reasonable to suppose that in the case of hydrogen the Sharp and Diffuse series *practically coincide*, the Balmer series representing the superposition of the two. There is not at present any direct evidence bearing on this question, but the above manner of regarding the Balmer series has the advantage of *bringing the hydrogen spectrum into line with other series spectra, and converting it from an exception into a limiting case.* (Curtis 1914, 619; emphasis added)

Thus, both Curtis and Sommerfeld sought to systematize their observations regarding the spectral lines of the elements in such a way that hydrogen would not stand out as a singularity and so that it would, rather, 'be brought into line' with the other elements and its peculiarities explained with the same theoretical resources used for the other elements.[11]

In his theoretical treatment, in addition to Bohr's quantum number n quantizing the size of circular orbits, Sommerfeld introduced another quantum number (referred to him as n') for the quantization of the 'shape' of orbits, or, more precisely, the eccentricity of elliptic orbits. By way of phase integrals, Sommerfeld was then able to relate *both* quantum numbers to the eccentricity of elliptic orbits, as described by the following formula:

---

[11] There were further precursors. The attempt to integrate the hydrogen lines into the lines of other elements has been tracked historically by Robotti (1983). Curiously, neither Curtis nor Sommerfeld play any part in Robotti's story.

$$1 - \varepsilon^2 = \frac{n^2}{(n + n')^2}$$

This allowed him to derive an expression for the energy of the system so that 'the energy is thus uniquely determined by the sum of the quanta of action [n + n'], which we may distribute arbitrarily among the azimuthal and radial coordinates' (Sommerfeld 1916a, 19). Sommerfeld called this the result of 'penetrating certainty' (*schlagende Bestimmtheit*). As can be seen from the preceding equation, if the newly introduced quantum number n' would be zero, the eccentricity would turn zero too, rendering the electron orbit circular. And this possibility Sommerfeld did allow. Thus, rather than circular orbits replacing elliptic orbits in the Sommerfeld model, the former *complemented* the latter. What is more, in his treatment of different line intensities, Sommerfeld even went so far as rendering circular motion *more likely*:

> We have assumed that the *circular orbit is the most likely* and that the elliptic orbit is the more likely the bigger its eccentricity. (Sommerfeld 1916a, 63)

With this formula, Sommerfeld was now able to accommodate what he conceived of as the 'Sonderstellung' of hydrogen with the theoretical machinery that also accommodated the spectral lines of other elements. The amended Balmer series formula, containing two quantum numbers $v = N\left(\frac{1}{(n+n')^2} - \frac{1}{(m+m')^2}\right)$ for hydrogen, would 'reduce, in a sense, by accident', to the standard formula containing only one quantum number, as derived originally by Bohr, namely the one for *circular* orbits (20).[12] Sommerfeld concluded that the Balmer series could now 'appear in a new light', as they could now be viewed as being caused in a number of different ways (with the second quantum number n' for elliptic orbits taking different values without this translating into different spectral lines). This, for Sommerfeld, had 'deepened theoretical significance' in comparison to Bohr's theory (Sommerfeld 1923, 237).[13] At the same time, Sommerfeld pointed out that the 'different ways of production' ('Erzeugungsarten') coincided in a single Balmer line 'only approximately' and could be retrieved by means of including relativistic effects

---

[12] In this formula, *n, n'* describe the initial electron orbit and *m, m'* the final electron orbits (i.e., before and after a quantum 'jump' of the electron).

[13] The Stark effect was accounted for by the introduction of a third quantum number, quantizing the 'location of the orbit' relative to an external electric field. This was done successfully by Sommerfeld's student Epstein and by Schwarzschild. See, e.g., Kragh (2012, 154) for details.

(Sommerfeld 1916a, 25). Sommerfeld then dedicated the entire second part of his treatment to the fine structure of hydrogen, and in particular to the hydrogen doublet. Crucially, Sommerfeld was able to derive the correct fine structure constant for hydrogen, which has been described as 'perhaps the most remarkable numerical coincidence in the history of physics' (R. Kronig, cited in Pais 1991, 188). The historian Kragh has concluded that 'by some sort of historical magic, Sommerfeld managed in 1916 to get the correct formula from what turned out to be an utterly inadequate model' (Kragh 1985 84).[14]

4.2.2.2   Assessment

We can make two crucial observations regarding this stage of the development of the Bohr–Sommerfeld model. First, the changes made to the model are inconsistent with both novelty criteria. The motivations stated by Bohr and Sommerfeld for the change from circular to elliptic orbits by the scientists in question were clearly governed by the attempt to accommodate the fine structure of hydrogen and other elements. On the standard accounts of novel success, this change must therefore be considered ad hoc: the model was changed in order to accommodate the fine structure; i.e., the evidence was used in the construction of the model that accommodated that model. Second, the introduction of elliptic motion into the model can hardly be understood in terms of de-idealization. McMullin presents the introduction of circular orbits as some kind of physical necessity in the goal of de-idealizing the model. As we observed, however, circular motion still figures in Sommerfeld's development of Bohr's model. What is more, Sommerfeld deemed it even *more likely* in order to make sense of line intensities.

Again, as in the first stage of the development of the Bohr–Sommerfeld model, the apparent violations of the novelty criteria, and the idea of de-idealization not being able to accommodate the actual development by Sommerfeld, Sommerfeld's model got an overwhelmingly positive reception. Einstein, for example, in a letter to Sommerfeld in February 1916, called Sommerfeld's treatment 'a revelation', which 'delighted' him and convinced him that Sommerfeld was 'right' (Einstein, in Eckert and Märker 2000, 524–5). About half a year later, Einstein wrote to Sommerfeld in August 1916:

---

[14] For a philosophical discussion of this achievement see Vickers (2012).

Your investigation of spectra belongs among my most beautiful experiences in physics. Only through it do Bohr's ideas become completely convincing. (563)

Bohr, likewise, said this on 19 March 1916: 'I do not believe ever to have read anything with more joy than your beautiful work' (603). And H. A. Lorentz even commented to Sommerfeld on 14 February 1917, 'You have reached one of the most beautiful results in theoretical physics' (574).

Have we perhaps treated novelty defenders unfairly? Should we not rather ask whether the changes made to the model *ultimately* resulted in novel success, rather than whether the changes themselves can count as novel success? Indeed, Sommerfeld's model did occasion confirmations of novel predictions. In particular, Sommerfeld correctly predicted the fine structure of ionized helium as confirmed by Paschen (1916). Yet neither Einstein's nor Bohr's assessment could have been influenced by this confirmation: both were made before Paschen's results.[15] So whatever persuasive power novel success might have, it was not required for Einstein and Bohr to approve of Sommerfeld's changes. Furthermore, not only was the accuracy of Paschen's measurements challenged by new high-precision measurements by the German physicists E. J. Gehrcke and E. Lau in early 1920, but also, as Kragh (1985) has pointed out, any empirical test of Sommerfeld's model was problematic because it depended on rather arbitrary intensity rules, such as the one previously mentioned (73 and 93). Thus, the novel success that the Bohr–Sommerfeld model did generate stood on shaky grounds.

4.2.3   A Material De-Idealization: Electron Spin

The previous three stages are all examples for what McMullin considers 'formal' de-idealizations, which he defines as de-idealizations of a model or theory which leave out features in the theoretical representation of a target system that are *known to be relevant to the explanation of the target system.* The fourth stage of McMullin's discussion of the development of the Bohr–Sommerfeld model – the invocation of the concept of spin by G. Uhlenbeck and S. Goudsmit in 1925 – is an instance of what

---

[15] On 21 May 1916, Friedrich Paschen informed Sommerfeld that he could experimentally confirm the predictions of Sommerfeld's model about the fine structure to be expected in ionized helium. Paschen reported the results to *Annalen der Physik* in July 1916, where they, according to the publisher (personal communication), saw print probably in September of the same year.

McMullin calls 'material' de-idealization. In that kind of de-idealization, one adds features to the theoretical representation of the target system which *were not known to be relevant* in the explanation of the target system when the first model or theory was devised (McMullin 1985, 258).[16]

4.2.3.1 *The Facts*[17]

The introduction of a third quantum number had allowed Sommerfeld (with the help of his assistant Epstein) to account not only for the Stark effect, but also for the so-called normal Zeeman effect, i.e., the splitting of lines in in the presence of a magnetic field. This effect had been known since 1897 and had already been explained by Lorentz in terms of his electron model, a feat for which he, together with Zeeman, received the Nobel Prize. In the same year, however, it was discovered that there were forms of the Zeeman effect, specifically in sodium and generally in elements producing multiplet spectral lines, which did not conform to Lorentz's predictions. For that reason, it came to be known as the 'anomalous Zeeman effect' (and is to this day). In fact, the anomalous Zeeman effect constituted a major problem also for the Bohr–Sommerfeld theory. In order to accommodate it, Sommerfeld introduced yet another quantum number in 1920. In contrast to the other three quantum numbers, Sommerfeld found no electron-orbit correlate for it. Instead, he attributed it to a somewhat-mysterious 'hidden rotation', which, due to the works of A. Landé and Heisenberg around 1921–1922, had come to be viewed as an angular momentum of the inner 'core' or 'rump' of electrons whereby the new quantum number quantized the states of (magnetic) interactions between that core and the single outer, valence, electron. With the help of this model and the introduction of half-integer quantum numbers, Landé, in 1921, managed to capture the line-splitting data on the alkali metals, but only at the cost of counting the contribution of the rump to the energy of the atom, somewhat arbitrarily, *twice*. Pauli showed that Landé's solution, although empirically adequate, led to absurd consequences (cf., e.g., Jammer 1989, 137). Instead, Pauli proposed that the 'hidden rotation' is not to be found in the core, as assumed by Landé, Heisenberg, and others, but rather in the valence electron. The anomalous Zeeman effect,

---

[16] Both formal and material idealizations are forms of construct idealizations, where simplification is a property of a 'conceptual representation of the object', in contrast to causal idealizations, where the 'problem situation itself is simplified, so that 'the diversity of causes found in Nature is reduced and made manageable' (McMullin 1985, 255 and 265).

[17] This section is based on the extensive historical discussions that can be found in Forman (1968), Mehra and Rechenberg (1982), Jammer (1989), Pais (1991), and Kragh (2012).

accordingly, would then be due to a classically non-describable 'two-valuedness' (German: *Zweideutigkeit*) of the electron. Extrapolating this idea from the context of alkalis in 1924, Pauli assumed that his new-found Zweideutigkeit was a property not just of electrons in alkali atoms, but also of electrons in any atom.[18]

In an attempt to return to the classical visualizability of the Bohr–Sommerfeld model, in line with Sommerfeld's interpretation of quantum numbers as corresponding to degrees of freedom of the electron, Uhlenbeck and Goudsmit in 1925, and R. Kronig independently of them, interpreted Pauli's Zweideutigkeit as the literal spinning of electrons around their own axis on their orbits around the nucleus.[19] The concept of electron spin explained not only the fine structure of hydrogen and the alkali elements more accurately than it was possible with the Bohr–Sommerfeld model, but it also accounted for the anomalous Zeeman effect. A discrepancy of a factor of 2 between the predictions derived from the concept of spin and the fine structure of hydrogen was removed by L. Thomas in 1926, who corrected a mistake in Uhlenbeck and Goudsmit's relativistic treatment of the electron. Another problem associated with the latter, however, remained: the absurd consequence of the surface of spinning electrons being faster than the speed of light, resulting from assigning electrons an extension, which was needed for the classical spin property. This issue remained unresolved until Pauli's development of spin matrices within the framework of a non-relativistic quantum mechanics in 1926.[20]

4.2.3.2 Assessment

The concept of electron spin, as put forward by Uhlenbeck and Goudsmit, was introduced in order to accommodate the Zeemann effect. Both the temporally novel and weak use-novel success criteria are clearly violated. As to the strong use-novel criterion, there was no free parameter in the Bohr model. Insofar as it makes sense to talk about parameters here in the first place, a free parameter was added to the model in order to be able to accommodate the data. Can the introduction of spin be understood as some kind of de-idealization?

---

[18] On the basis of this assumption, Pauli formulated the famous exclusion principle. See the cited sources for details.

[19] See de Regt (2001) for a nice discussion of the role of visualizability in physics.

[20] Non-relativistic quantum mechanics turned out to be a limit of Dirac's relativistic equation of the electron.

It is interesting to note, first of all, that McMullin and Lakatos disagree about whether or not the concept of spin was suggested already by the original model by Bohr. Whereas Lakatos believes that spin was 'planned right at the start' alongside a finite mass nucleus and elliptic orbits (1970a, 146), McMullin has it that in the original model, 'there was no reason to suppose that the electron would possess spin' (McMullin 1985, 263).[21] It is questionable whether one can arbitrate in a meaningful way here, save for pointing out that it would be strange if spin had been suggested by the original model and then not been considered for more than 10 years before Uhlenbeck and Goudsmit introduced the idea in 1925.

In spite of his reluctance to consider spin as being suggested by the original model, McMullin does believe that the addition of spin ought to count in favour of the fertility of the Bohr–Sommerfeld's model. This is slightly strange given that McMullin in his account stipulates de-idealization as a necessary condition for changes not to be ad hoc. Still, he concludes in favour of the fertility of the Bohr–Sommerfeld model by saying that 'from this simple physical assumption the correct answers were found for both Zeeman effects, not only in hydrogen but in other types of atom as well' (McMullin 1968, 395).

In stark contrast to McMullin's positive assessment of the introduction of spin being a fertile modification, Lakatos concludes his brief remarks about the introduction of spin by saying that 'the temerity in proposing wild inconsistencies did not reap any more rewards. The programme lagged behind the discovery of "facts"' (154). That is, Lakatos denigrates the change on the basis of it not ultimately resulting in novel success, which, as we saw previously, for Lakatos amounts to the degeneration of a research programme. Yet, once again, such retrospective justification seemed not to be required for contemporary physicists to view the Sommerfeld model favourably. Bohr, for example, in an attached note to Uhlenbeck and Goudsmit's *Nature* article (which appeared in 1926, i.e., one year after its German version), wrote that the concept of spin 'promises to be a very welcome supplement to our ideas of atomic structure' (264). Around the same time, in a letter to Kronig, Bohr described his ultimate conversion to Uhlenbeck and Goudsmit's idea: 'I have never since faltered in my conviction that we at last were at the end

---

[21] Lakatos appears to suggest that Bohr himself already had that modification in mind when designing the original model (thanks to Helge Kragh for pointing this out to me). There is no textual evidence that Bohr did, but in any case, this seems not to matter so much for whether or not the original Bohr *theory*, or *programme*, bore out that suggestion.

of our sorrows' (Pais 1991, 243).[22] Also, Pauli, who had earlier called spin an 'Irrlehre' (mistaken idea), had conceded that 'although at first I strongly doubted the correctness of this idea because of its classical mechanical character, I was finally converted to it by [L. H.] Thomas's calculations on the magnitude of doublet splitting . . .' (Pauli's Nobel Prize lecture, 13 December 1946). Indeed, despite the aforementioned absurd consequence of electron surfaces spinning faster than the speed of light (in violation of the special theory of relativity), Pauli was happy to declare in a letter to Bohr on 12 March 1926 that '[n]ow nothing else is left for me than to surrender completely!' (Pauli 1979, 310). Although Heisenberg also was initially sceptical towards spin and sought to accommodate the hydrogen fine structure with his rump model (see earlier), he ultimately admitted that 'naturally, it is an enormous advance that your theory harmoniously explains away all sorts of things like *Zwang* and the rest' (Heisenberg to Goudsmit, 16 December 1925, cited by Serwer 1977, 251). Again, these reactions were anything *but* negative, regardless of any possible future novel success resulting from the introduction of spin. Indeed, soon after its introduction, the history of the old quantum mechanics came to an end and classical concepts were given up entirely.

4.2.3.3 The Bohr–Sommerfeld Model Concluded

Many changes of the Bohr–Sommerfeld model were made in order to accommodate anomalies. According to the standard accounts of novel success, that makes these changes ad hoc. This is not the case, however, on McMullin's account. Changes to the model in response to anomalies are not ad hoc so long as they are warranted by de-idealizations of the model. As it turns out, though, not even McMullin's preferred example supports this idea. Sommerfeld was happy to keep circular motion in his model and even made it the *most likely* motion in order to accommodate line intensities. There is also no indication in the publications of either Bohr or Sommerfeld that they considered going elliptic as some kind of physically necessitated de-idealization. Bohr, for example, toyed with the idea of introducing elliptic orbits *only* when trying to accommodate the fine structure of the spectral lines of hydrogen (cf. Section 4.2.2). Although it is hard to see that de-idealization could justify the changes of the Bohr model, there seems to be *something* about the Bohr model that allows for

---

[22] Bohr's conversion happened after he learned from Einstein (through Ehrenfest) that the right application of the theory of special relativity could produce the magnetic field needed for electron spin. See Pais (1991, 243).

non–ad hoc changes in the accommodation of anomalous evidence.[23] This is by no means guaranteed. Sometimes theories can accommodate new evidence only in an ad hoc fashion. To see this, let us consider briefly another historical example, namely the specific heat anomaly that challenged the kinetic theory of heat in the second half of the twentieth century.[24]

4.3   An Objectionable Change

The kinetic theory of gases conceives of heat as molecular motion. In its most primitive form, gas particles are assumed to be perfectly elastic and of negligible size. In 1857, one of the pioneers of the kinetic theory, the German physicist Clausius, derived the first predictions for the specific heat capacity of gases, which corresponds to how much energy is required to raise a gas's temperature by one degree. Assuming the equal distribution of energy along a molecule's three translational directions in space, the prediction was $\gamma = \frac{5}{3} = 1.66$, given the formula $\gamma = \frac{c_p}{c_v} = \frac{2+n}{n}$ (whereby c_p and c_v = specific heat capacity at constant pressure and volume, respectively, and n = the number of degrees of freedom). Unfortunately, measurements of the specific heat ratios of oxygen, nitrogen, and hydrogen gave only $\gamma = 1.42$ and thus contradicted this prediction. Maxwell, who did much to develop the kinetic theory, was the first to fully appreciate this problem, which came to be known as the 'specific heat anomaly'. In fact, Maxwell believed that the anomaly 'overturns the whole hypothesis [of the kinetic theory of gases], however satisfactory the other results may be' (Maxwell in 1860, cited by Brush 1976, 194).

Around the same time, chemists and physicists started to realize that most elements are diatomic in the gaseous state. Diatomic molecules, however, would add another two degrees of freedom, namely two degrees of rotation. Furthermore, the known spectral line emission data for various gases suggested that gas molecules are extended structures that would allow for internal, periodic vibration. But an additional, vibratory degree of freedom, which Maxwell considered in 1875, would not result in the right prediction either: the three translational degrees of freedom plus

---

[23] Note that we need not believe that all of the changes to the Bohr model were non–ad hoc. See Section 4.2.2.2.

[24] Excellent historical and philosophical discussions can be found in Brush (1976), Clark (1976), Nyhof (1988), and de Regt (1996).

one vibratory and two rotational degrees of freedom yielded $\gamma = \frac{8}{6} = 1.33$ instead of the measured 1.4. Luckily for the kinetic theory, there was finally some good news in 1976: Kundt and Warburg had discovered that monoatomic mercury vapour had a specific heat of 1.66, i.e., the specific heat predicted for gases with just rotational degrees of freedom. Inspired by this discovery, Boltzmann, in the same year, simply stipulated in an ad hoc fashion that molecules with a specific heat ration of 1.4 were diatomic and behaved like dumb-bells: they were rigidly connected and therefore could not vibrate, resulting in $\gamma = \frac{7}{5} = 1.4$, in accordance with the measurements.

Although Maxwell conceded that Boltzmann's proposal was 'in striking agreement with the phenomena of the three groups of gases [i.e., monoatomic, diatomic, and polyatomic]', and although he called Boltzmann's idea a 'somewhat promising hypothesis', Maxwell strongly objected to it on the grounds of it lacking physical plausibility (Maxwell 1860, cited in Brush 1976, 194). In particular, he objected to Boltzmann's dumb-bell model not allowing for internal vibrations when, Maxwell was certain, spectral line emission data required them (194). This sentiment was shared amongst other physicists. For example, Lord Kelvin believed that it was 'rigidly demonstrable that repeated mutual impact must gradually convert all the translational energy into the energy of shriller and shriller vibrations of the molecule' (1886, cited in Clark 1976, 85). Lord Rayleigh commented in 1900:

> However great may be the energy required to alter the distance of the two atoms in a diatomic molecule, practical rigidity is never secured, and the kinetic energy of the relative motion in the line of junction is the same as if the tie were of the feeblest. The two atoms, however related, remain two atoms, and the degrees of freedom remain six [not five as in Boltzmann's proposal] in number. (cited in Clark 1976)

Only quantum mechanics would eventually bear out Boltzmann.[25]

The contrast between Boltzmann's accommodation of the specific heat anomaly and the changes made to the Bohr model is striking. Although in both cases the model/theory was changed in response to anomalies it was facing, and although in both cases a match with the phenomena was established, only in the case of the Bohr model were those changes deemed permissible. This is particularly interesting given that (i) the Bohr model

---

[25] According to quantum mechanics, vibrations become significant only at high temperatures (Nyhof 1988, 99).

was tapping into new physical territory where accepted physics had reached its limits and given that (ii) changes to the Bohr model also resulted in apparently absurd consequences, notably after the introduction of spin. Again, although physically plausible (or even necessary) de-idealization does not seem to be the right way of thinking about permitted changes, there does seem to be something about the Bohr model that allows new evidence to be accommodated in a non–ad hoc fashion. In contrast, the kinetic theory, with regard to the specific heat anomaly, lacked this capacity.

4.4   Conclusion

In this chapter I critically assessed McMulllin's construal of theoretical fertility. Instead of thinking of fertility in terms of novel success, which has figured so prominently in the realism debate of recent years, McMullin conceives of fertility in terms of the successful (i.e., non–ad hoc) accommodation of anomalies by way of de-idealization. McMullin believes that this kind of fertility gives us strong (and perhaps the strongest) grounds for realism. As I showed in this chapter, though, McMullin's preferred example, the Bohr–Sommerfeld model of the atom, does not support this idea. Deidealization simply was not part of the agenda of scientists making changes to the model. Also, the changes were always made only *after* anomalies were discovered, which is curious: a proper de-idealization of a model does not need to await 'triggering' through an anomaly, but is driven by model-internal considerations (Lakatos fully embraces this idea). Most importantly, perhaps, several changes did not go the way they should have, if they were simply de-idealizations (e.g., Sommerfeld not only retaining circular orbits, but also rendering them the most likely electron motion).

What seems right about McMullin's treatment of the case is that *something* about the model allowed it to accommodate the anomalies in a non–ad hoc fashion, even if that 'something' has nothing to do with de-idealization. This is particularly apparent when the Bohr–Sommerfeld model is compared to the kinetic theory of gases with regard to the specific heat anomaly: the kinetic theory could not accommodate the specific heat of diatomic molecules without Boltzmann's ad hoc modification. But what precisely made it ad hoc? What are the general conditions of ad hocness, if there are any? This is the topic of our next chapter.

CHAPTER 5

Ad Hoc Hypotheses and the Argument from Coherence

As we saw in Chapters 3 and 4, the rationales for both novel success and M-fertility are based on the desire to block the possibility of ad hoc manoeuvres. In this chapter we want to develop a better understanding of what it means for a hypothesis to be ad hoc in the first place.

Intuitively, a hypothesis is ad hoc, minimally, when it is introduced for the sole purpose of 'saving' a theory which faces observational anomalies. Whilst this definition may suffice for all practical purposes, it is not enough for anybody looking for an *epistemic* account. What is it about a hypothesis that somehow makes it less deserving of confirmation than a hypothesis that accommodates the same evidence in a non–ad hoc fashion? The foregoing intuitive definition gives us no answer to this question. It only tells us something about *why* ad hoc hypotheses are used; it tells us about motivations. It does not tell us anything about the reason of epistemic deficiency.

It goes without saying that a good account of ad hoc judgements in science must take into consideration how scientists actually use ad hoc judgements. Traditionally, discussions of ad hocness have been informed by historical case studies. The present chapter will follow this tradition, although there might be other ways of investigating the meaning of ad hocness by descriptive means (interviews with scientists?). But merely describing the circumstances under which scientists deem hypotheses ad hoc will not suffice either: scientists generally do not explicate the grounds of their judgements; they just make them. Also, when assessing hypotheses, they may not always be very explicit and may not use the word 'ad hoc' but perhaps other normative language such as 'artificial', 'concocted', and 'contrived'. Thus, some substantive philosophical work is indispensable in the project of determining the meaning of ad hocness by descriptive means.

The structure of this chapter is as follows. In Section 5.1, I will lay out the state of the art of regarding philosophical accounts of ad hocness.

In Section 5.2 I propose my own view of ad hocness and explain its conceptual and descriptive attractions. Section 5.3 gives further in-depth illustrations. On the basis of these consideration I will, in Section 5.4, propose my second virtuous argument for realism.

5.1   Ad Hocness: The State of the Art

This section critically assesses five accounts: ad hocness as a lack of independent testability (5.1.1) or independent support (5.1.2), as a lack of unifiedness (5.1.3), as the fixing of free parameters (5.1.4), and as subjectivist/aesthetic projections (5.1.5). Although I do not believe that any of these accounts is ultimately refuted by my discussion, I do think that the standard accounts face some serious challenges, some of which have not been brought forward strongly enough.

5.1.1   Ad Hocness as Lack of Independent Testability

As we mentioned in Chapter 1, Popper, being well aware that no experiment could force us to refute a theory, emphasizes that theories ought not to be saved by ad hoc manoeuvres. In fact, his falsificationism was once even described as 'introducing new, non-justificationist criteria for appraising scientific theories based on anti-ad hocness' (cf. Popper 1959a, 42; Lakatos 1978, 39). Popper's answer to the question of 'Why are ad hoc hypotheses deficient?' was that that they decrease the falsifiability of the theory they were introduced to save (Popper 1959a, 82f.) and that 'degrees of ad hocness are related (inversely) to degrees of testability and significance' (Popper 1959b). Popper's reasoning appears to be that, if a hypothesis which is added to a theory entails only the evidence that is troublesome to this theory, then the falsifiability of the theory is decreased: there is one state of affairs less that could threaten the theory in question. Contrapositively, the falsifiability of the theory is increased when the added hypothesis makes new predictions. This is why Popper also claimed that ad hoc hypotheses 'cannot be tested independently' (Popper 1976, 986), i.e., they cannot be tested other than on the basis of the evidence they were introduced to account for. At least this is the idea.

Barnes (2008, 11) has, following Bamford (1993, 349–50) – I think rightly – questioned Popper's alleged connection between ad hoc hypotheses and decreased testability. It is, for example, not plausible that the hypothesis 'bread nourishes' when amended to 'all bread nourishes except that grown in a particular region of France' would, after the relevant discovery, result in a decrease in testability. In fact, those two sentences seem equally testable, with the only difference being that bread does or does not nourish in a particular region in France. Indeed, the hypothesis is even independently testable: one can go and find out (empirically) about the particular reasons why the bread in that particular region in France does not nourish, for example, by investigating the wheat used in the process, etc.

Independent testability also seems not the right way of thinking about one of the most emblematic ad hoc hypotheses, namely the Lorentz–FitzGerald contraction hypothesis (LFC), famously introduced in 1892 to save the ether theory from refutation in the face of the famous Michelson and Morley ether drift null result from 1887. There is strong evidence that physicists at the time regarded the LFC as an ad hoc hypothesis. Einstein, for instance, wrote about LFC that '[t]his manner of theoretically trying to do justice to experiments with negative results through ad hoc contrived hypotheses is highly unsatisfactory' (Einstein, in Warburg 1915, 707). To Michelson, who initially was actually quite keen to explain away his null result in favour of the ether theory, 'such a hypothesis seems rather artificial' (cited in Holton 1969, 139). Even Lorentz, who had developed the LFC, admitted that it first appeared 'far-fetched' and that it depended on an assumption (namely that the same laws apply to inter-molecular forces that apply to electrical ones) which 'there is no reason to make' (Lorentz 1895, in Einstein et al. 1952, 6). In response to criticism by Poincare in 1900, Lorentz also admitted that '[surely] this course of inventing special hypotheses for each new experimental result is somewhat artificial' (Lorentz 1904, in Einstein et al. 1952, 12–13).

Although Popper, in his *Logic of Scientific Discovery*, mentioned the LFC as the only example of his claim that ad hoc hypotheses are not indepen-dently testable, Grünbaum (1959) pointed out that LFC actually *did* entail an independently testable prediction. There was even an experiment, namely the so-called Kennedy–Thorndike experiment, which falsified this consequence of the LFC.[1] In response, Popper conceded that the LFC could not be disqualified as ad hoc on the basis of it not being independently testable.[2] The conclusion Grünbaum (1976) drew was

---

[1] The LFC had it that matter moving through the ether contracts in the direction of motion by the same laws governing the contraction of interference patterns. Whereas the Michelson–Morley experiment showed that the speed of light is independent of the orientation of the experimental apparatus, the Kennedy–Thorndike experiment showed that it was independent also of the velocity of the apparatus. For more details see, e.g., Janssen (2002b).

[2] Popper nevertheless held onto the independent testability account and claimed that the LFC was at least less independently testable (and therefore more ad hoc) than Einstein's special theory of relativity (which he considered ad hoc to some degree also) (Popper 1959b).

this: ad hoc hypotheses *can* have independent consequences, but when they do, these consequences either remain without support (because no relevant evidence can be obtained) or they are shown to be false.

5.1.2   *Ad Hocness as the Lack of Independent Support*

The idea that a hypothesis needs independent support for it not to be judged ad hoc is extremely popular amongst philosophers (e.g., Zahar 1973a; Schaffner 1974; Leplin 1975; Lakatos 1978; Scerri and Worrall 2001; Worrall 2002; Sober 2008).[3] Yet there are several reasons to be wary of this notion.

Consider an example made by Worrall (2002) in the context of predictivism. Creationists, when presented with fossils seemingly undermining the creationist doctrine, might respond with a 'Gosse dodge': fossils are just God's playful writings in stone, possibly testing our faith. Worrall suggests that the Gosse dodge is ad hoc because it has no independent support. However, suppose now with Barnes (2005) that we view the Gosse dodge as a general hypothesis about *all* fossils. Then the Gosse dodge would not only be about the particular fossils for which the Gosse dodge has been invented, but also about *other* fossils throughout the world and fossils yet to be discovered. Arguably, those other fossils, once discovered, would be independent support for the Gosse dodge. Yet the Gosse dodge is clearly ad hoc, regardless. One may not be impressed with this toy example; let us therefore turn again to historical examples.[4]

Proponents of the independent support view of ad hocness have repeatedly appealed to the discovery of the planet Neptune to make their case (Worrall 2002). In order to accommodate a discrepancy in the orbit of Uranus, Adams and Le Verrier, independently from each other, postulated

---

[3] Lakatos distinguished between three senses of ad hoc hypotheses (added to a research programme): ad hoc$_1$ = the hypothesis makes no new predictions; ad hoc$_2$ = the predictions made by a hypothesis are not confirmed; ad hoc$_3$ = the hypothesis does 'not form an integral part of the positive heuristics' (Lakatos 1978, 112, fn. 1). The first and the second sense are covered by my discussion in this and the previous section. A discussion of the third sense here would lead us too far astray.

[4] One might, for example, object to this example on the basis of (i) the vagueness of the Gosse dodge and (ii) the insufficiency of independence of the support from other fossils. As to the first point, one may make the criticism that the Gosse dodge is not really falsifiable and therefore should not be taken seriously as a scientific hypothesis in the first place. As to the second point, one may demand, on behalf of the proponents of the independent support account, that empirical support must be qualitatively significantly different for it to be independent. I sympathize with both points. Yet I do share the crucial intuition that independent support for a hypothesis is not sufficient for rendering it non–ad hoc.

another planet in the vicinity of Uranus in 1845–1846. Neptune was subsequently discovered in 1846. Proponents of the independent support view of ad hocness have taken this to confirm their view (Worrall 2002): the Neptune hypothesis was introduced ad hoc in order to save Newtonian mechanics from refutation but was rendered non–ad hoc when it received independent confirmation via the discovery of Neptune. However, as Leplin (1982) has pointed out, there is *no evidence whatsoever* that the scientific community at the time regarded the postulation of Neptune ad hoc, or somehow methodologically unsound.[5] But if the postulation and discovery of Neptune really is no example of an ad hoc hypothesis, then it cannot lend credence to the independent support condition of ad hocness, or to *any* account of ad hocness for that matter. Moreover, it can then also not be the case that introducing a hypothesis to save a theory from refutation is sufficient for that hypothesis to be ad hoc. Something else is needed.

With regard to the aforementioned contraction hypothesis, one may also note with the historian Holton that the question of independent support appeared not to matter much in its assessment, since its independent predictions 'were not urged as tests that would decide on its acceptability' (Holton 1969, 177). Incidentally, it is interesting to note with Leplin (1975, 314, fn. 17) that Lorentz appears to have regarded the question of whether or not the LFC was ad hoc independently of the question of whether or not there ever could be (positive) experimental tests for it (Lorentz 1885, in Einstein et al. 1952, 6). We shall return to this point later in this chapter (Section 5.3.1).

The independent support view of ad hocness is very persistent. In the most recent contribution on the topic of ad hocness, Friederich et al. (2014) discuss several judgements by physicists (and some philosophers) about the Higgs mechanism (HM) of the standard model in particle physics being ad hoc. Friederich et al. list the following features of the HM that arguably underlie physicists' ad hoc judgements: (i) the HM leads to a large number of free parameters for the particle masses that are not determined in a principled fashion but need to be put in 'by hand' on the basis of experimental results; (ii) there are no known fundamental scalar particles in physics apart from the Higgs boson; (iii) the symmetry-breaking of the HM, *contrary to all other known cases of symmetry-breaking*, is implemented

---

[5] See, for example, Grant (1852, 164–201) and Grosser (1962). Bamford (1996), similarly, notes that 'here was no theoretical objection, however, to a planet located at some intermediate distance beyond Uranus' (216).

not dynamically but by fiat;[6] and (iv) the fact that the bare mass of the Higgs particle is 'fine-tuned' to its interaction mass in an 'unnatural way' (by 34 orders of magnitude) is unexplained.[7] This is known as the 'fine-tuning' or 'naturalness' problem and, according to Friederich et al., 'is widely regarded as the most severe' argument against the 'fundamental scalar character' of the Higgs field.[8] In other words, the HM has a number of arbitrary, theoretically unmotivated features. Interestingly, this is also what Weinberg expressed in a paper which laid the foundation for the standard model. There he conceded that 'our model has too many arbitrary features for these predictions [regarding the electroweak "mixing angle"][9] to be taken very seriously' (Weinberg 1967, 1265–6).

Friederich et al. argue that, despite these early concerns, HM is no longer ad hoc, because 'the most crucial characteristic of an ad hoc hypothesis', which for them is the lack of independent support, 'is no longer obeyed' (3913).[10] Yet at the same time, they remark that

> [m]any physicists ... now seem to be ready to accept the [HM] as part of physical reality. On the other hand, most of them seem to be *not* ready to conclude that the [aforementioned] criticisms of the [HM] were unfounded ... [since] the experimental confirmation of the [HM] does not solve any of these problems. (3913; original emphasis)

But if the reasons for judging the HM ad hoc have not been addressed by the Higgs discovery, then why should we think that the ad hoc status of the HM has changed? The fact that the HM describes a particle that exists is

---

[6] Symmetry-breaking is normally due to composite rather than fundamental fields. Only in the former case can it be dynamical.

[7] In quantum field theory, any experimentally determined particle mass is understood as the sum of the 'bare' mass and the 'interaction mass', which is the mass of a particle interacting with vacuum fluctuations.

[8] Friederich et al. mention two further points: (v) that a non-zero Higgs field in a vacuum is as conceptually problematic as the aether was, and (vi) 'triviality'. The first of these points, Friederich et al. argue, makes a false presupposition: the Higgs field in the vacuum is no physical quantity. The second concerns the position of the so-called Landau pole, basically a limit of the standard model to certain energy scales. The Landau pole for the standard model depends on the observed Higgs mass. For the currently observed Higgs mass, the Landau pole turns out to be entirely absent. This result, Friederich et al. point out, can be generated only 'if the self-coupling of the Higgs boson is assumed to be vanishing, i.e., trivial, while this self-coupling must be non-vanishing for the HM to generate non-vanishing particle masses'. For reasons of simplicity, I will leave out those points in my discussion.

[9] The mixing angle of the Weinberg model, which also became known as the Weinberg angle, indicates the relative strengths of the neutral and charged weak interactions and the masses of the Z and W bosons.

[10] Instead of HM, Friederich et al. use the abbreviation SMHM for Standard Model Higgs Mechanism.

not enough. In the history of science, there have been many ad hoc hypotheses of real entities or phenomena (such as the Planck hypothesis of black-body radiation; see Section 5.2.1). Yet Friederich et al. seem to implicitly rule out this option, as they go on to suggest only two possible reactions to the Higgs discovery: either supersymmetry theories, which incorporate the Higgs boson as one amongst several such particles, may eventually turn out to be correct or the standard model is confirmed by the Higgs discovery and the HM will therefore no longer be viewed as ad hoc.[11] But, again, I think this is a false dichotomy: it may well be that HM is and *will remain* ad hoc despite the discovery of the Higgs particle, until a better theoretical device in agreement with the facts has been constructed.[12] Since none of HM's original problems listed by Friederich et al. has disappeared (which led to the ad hoc judgements), this seems to be the more accurate picture.

In sum, the independent support view of ad hocness is in trouble. Physicists' view of the HM, the (hypothetical) Gosse dodge, and (I would argue) the LFC show that a hypothesis can be ad hoc even when it does (or if it would) have independent support. The Neptune hypothesis – often used as simple illustration for the independent support account – was never deemed ad hoc, even at a time when it had no independent support. The lack of independent support thus seems neither necessary nor sufficient for a hypothesis to be considered ad hoc.

---

[11] Friederich et al. offer quotes from three physicists commenting on two of the problems of the HM as evidence for the second option. But none of these quotes seems to support this view. Friederich et al. refer to a review article on the status of supersymmetry by Feng (2013), a blog entry by Krämer (2013), and a paper on the hierarchy problem by Wetterich (2012). Feng concludes his paper by saying that 'weak-scale supersymmetry [motivated by the problem of naturalness, amongst other things] is neither unscathed, nor is it dead', which would seem to suggest that, despite the Higgs discovery, he has not given up the hope that naturalness will eventually be achieved in supersymmetric theories. There certainly is no evidence that he would regard the HM as less ad hoc than before the Higgs discovery. The same is true of Friederich et al.'s quote of Krämer: Krämer says only that the lack of evidence for supersymmetry in the results of the Large Hadron Collider experiments means that the search for theories incorporating naturalness may have to be given up and *not* that theories without naturalness, such as the HM, are now fully embraced as non–ad hoc because the Higgs has been found (as Friederich et al. seem to want to suggest). Lastly, it is worth noting that the quote by Wetterich, which Friederich et al. use to argue for a re-evaluation of the fine-tuning problem due to the Higgs discovery, is truncated and in fact starts with 'It has been shown long ago . . .', where Wetterich cites an article of his from 1984, i.e., 28 years (!) before the discovery of the Higgs particle.

[12] Physicists are indeed keen to develop such an alternative. It is, for example, well known that many physicists were rather disappointed that no further Higgs-like bosons were found at the LHC at CERN, as required by supersymmetry theories. See, e.g., Overbye (2012) and Heilprin (2013). String theory research, despite its problems, also still holds the promise of a theoretically more appealing theory than the standard model.

5.1.3   Ad Hocness as Lack of Unifiedness

One of the most systematic and most influential accounts of ad hocness is the one by Leplin (1975). Leplin's account is complex. It contains no less than five detailed, individually necessary, and jointly sufficient conditions for ad hocness (336–7):

1. A hypothesis H introduced into a theory T in response to an experimental result E is ad hoc if and only if
   a. E is anomalous for T but not for T as supplemented by H.
2. E is evidence for H, but
   a. no available experimental results other than E support H.
   b. H has no application to the domain of T apart from E.
   c. H has no independent theoretical support.
3. There are sufficient grounds neither for holding that H is true nor for holding that H is false.
4. H is consistent with accepted theory and with the essential propositions of T.
5. There are problems other than E confronting T which there is good reason to hold are connected with E in the following respects:
   a. These problems together with E indicate that T is non-fundamental.
   b. None of these problems including E can be satisfactorily solved unless this non-fundamentality is removed.
   c. A satisfactory solution to any of these problems including E must contribute to the solution of the others.

According to Leplin's condition 1 – i.e., the *condition of experimental anomaly* – ad hoc hypotheses in science are introduced in response to an experimental anomaly. Interestingly, though, Leplin himself allows for this condition to be removed should cases of ad hocness be found that suggest the inadequacy of this condition (1975, 336).[13] Condition 2, Leplin's *condition of justification*, captures the intuition that ad hoc hypotheses have no other support than the one they are designed to accommodate. Condition 3 is Leplin's *condition of tentativeness*. Presumably, more evidence than E is needed for judging whether H is correct. The condition of tentativeness is thus closely related to the condition of justification: if there were more evidence than E, we wouldn't have to be tentative about H. Condition 4, the *condition of consistency*, must be satisfied, because otherwise the

---

[13] Friederich et al. (2014) claim they have found such a case in the Higgs mechanism.

hypothesis in question would presumably simply be dismissed and not considered a serious candidate for amending T in the first place. Condition 5 is Leplin's *condition of non-fundamentality*. It is probably the condition most unique to Leplin's account. For Leplin, a theory is non-fundamental 'if no satisfactory solution of [the anomalous phenomenon] P can be achieved without the rejection of propositions in [theory] T and their replacement by propositions inconsistent with T' (325). What Leplin seems to suggest is that the theory T that is being amended with a hypothesis H in order to accommodate P lacks the resources for *unifying* the set of interlinked problems that P is a part of. Thus, in Leplin's view, the ad hoc charge is not so much directed against H but rather against T that H ought to save. Leplin motivates this view by referring to the criticism of the Lorentz–FitzGerald contraction hypothesis particularly by Einstein as 'directed primarily at the theory, and only indirectly at the particular hypotheses proposed as supplementation' (320).

There is an immediate problem with Leplin's account, relating to condition 5: if there is a phenomenon E that the theory in question accommodates by invoking a hypothesis H – in a seemingly ad hoc fashion – then H cannot be ad hoc when there are no phenomena other than E that indicate the non-fundamentality of T. But that is highly counterintuitive. A plausible account of ad hocness, I take it, must allow for the possibility of a hypothesis or theory to be ad hoc with regard to a single phenomenon. Also, there seems little reason to suppose a priori (as Leplin does) that every time a theory is considered ad hoc, the anomalous phenomenon can in fact be connected with other phenomena (by a more unified theory). This may be so in some cases, but it shouldn't be presumed for *all* cases.

The idea that ad hoc charges have got something to do with lacking unifiedness has currency also among other philosophers. Boudry and Leuridan (2011), in their criticism of Sober (2008), for example, claim that Paley's intelligent design hypothesis is objectionable not only because it has no independent support (as Sober claims), but also because it fails to unify the phenomena in a simple way (570). Lipton (1991/2004), too, can be said to hold a view that associates ad hocness with a lack of unification. For Lipton, when data are accommodated rather than predicted, 'there is a motive to force a theory and auxiliaries to make the accommodation. The scientist knows the answer she must get, and she does whatever it takes to get it. The result may be an unnatural choice or modification of the theory and auxiliaries that results in a relatively poor explanation and so weak support . . .' (170). In the case of prediction, in contrast, the scientist will 'make her prediction on the basis of the most natural and most explanatory theory and auxiliaries she can produce' (170). Whilst a scientist would not necessarily make an 'unnatural' modification of the theory when she accommodates (rather than predicts) data, according to Lipton, there is 'reason to believe' that a theory is being 'fudged' (as Lipton also likes to refer to 'unnatural' modifications) when it accommodates data. There is no such reason, according to Lipton, when the theory predicts the data. Such reasons have special prominence in Lipton's account, because, for Lipton, 'inductive support is translucent, not transparent' (178). Even the scientist proposing a theory 'should not assume that she is not doing [fudging] just because she is not aware that she is' (179).[14]

It's not really clear what Lipton means by a theory being 'natural' and, conversely, what he means by unnatural or 'fudged' theory modifications. He even admits himself that it is a 'clear limitation' of his account 'that it does not include anything like a precise characterization of the features that make one theoretical system fudgier than another' (180). He does, however briefly, mention that, for him, a fudged theory 'becomes more like an arbitrary conjunction, less like a unified theory' (171). If 'fudging' is just a different term for ad hocness (and I don't think that would be an overinterpretation), then Lipton can indeed be read as sympathizing with the unificationist account of ad hocness.

Harker (2008) is also sympathetic to the idea of the unification of the phenomena underlying the relative importance of predictive over accommodative success. More specifically, Harker claims that predictive success is apt to impress us only insofar as the theory in question, when successfully predicting a phenomenon E, at the same time also unifies E with other phenomena without having to invoke any further assumptions (Harker 2008). A theory's 'verified forecasts', as Harker also calls successful predictions, are of course also explananda of that theory. And what scientists, according to Harker, appreciate about verified forecasts is not so much the fact that the evidence in question was successfully predicted, but rather that the evidence was *explained* in a simple and unified way with the resources of the theory. For example, in the case of Fresnel's successful prediction of the famous white spot (Worrall 1989b), allegedly 'no additional assumptions were required' (Harker 2008, 447). Fresnel's theory, after this successful prediction, 'enjoys increased explanatory strength without any loss of

---

[14] Oddly enough, Lipton mentions the advance in the perihelion of Mercury as a prediction by Einstein's theory of 'special' [sic!] relativity (170). Usually that example is regarded as an example of an explanation rather than a prediction (Worrall 1989b, Brush 1994).

theoretical simplicity' (447). According to Harker, there is therefore no difference in epistemic import between accommodations and successful predictions as long as the phenomena are explained in a simple and unified way. Conversely, both accommodations and successful predictions will be of low epistemic quality if numerous assumptions have to be invoked for each phenomenon to be explained. Although Harker does not explicitly link the lack of unification to ad hocness (he is mainly interested in explaining the *apparent* appeal of predictive success), his account nevertheless suggests such a link.

I doubt that Harker's insistence on unifying theories not invoking 'additional assumptions' when incorporating new phenomena is sustainable. From Duhem we know that theories are never tested in isolation; tests of theories always depend on a host of auxiliary and background assumptions. And particularly in our attempts to confirm newly predicted phenomena, we will have to invoke a number of 'additional assumptions' when designing new experiments and using new instruments. Furthermore, whether or not, for example, Einstein's theory makes do with fewer assumptions than Lorentz's in accommodating the Michelson–Morley experiment (as Harker would require) is not obvious, I take it, and probably is not easily determinable. Yet that does not seem to matter to judgements about the LFC being ad hoc. I am therefore sceptical that operating with the *number* of assumptions will help us illuminate the notion of ad hocness (see also Section 5.2.1).

Finally, Lange (2001), as we saw in Chapter 3, defends a view according to which predictions count more than accommodations, because when evidence is accommodated, 'it is possible for the resulting hypothesis to be an *arbitrary conjunction*', whereas it is 'exceedingly unlikely' that an arbitrary conjunction is proposed before the relevant evidence is in (2001, in particular 583–4; emphasis added). Predictions are therefore prima facie more trustworthy than accommodations. With arbitrary conjuncts, '*apart from the evidence being accommodated*, there is no motivation for fastening onto that particular hypothesis rather than onto one with different conjuncts' (584; emphasis added). And because an arbitrary conjunct is 'prompted by little in the way of *theoretical* considerations', an arbitrary conjunct being true is 'likely to be utterly coincidental rather than to possess some physical significance' (2001; emphasis added). This is why, in Lange's view, for example, the Lorentz–FitzGerald contraction hypothesis received little support from the evidence it accommodated: 'not directly because Lorentz formulated the contraction hypothesis to *accommodate* the optical evidence ... [but rather because] the contraction hypothesis together with the rest of Lorentz's electrodynamics forms an arbitrary conjunction' (583).[15] Although Lange nowhere mentions ad hocness, I think it is safe to say that Lange associates arbitrary conjuncts with it. On the other hand, it's not so clear what the converse of arbitrary conjuncts is supposed to be. Lange has been read as placing epistemic weight on unifiedness (Harker 2008), although one may well interpret his writing in terms of coherence (see Section 5.2).

5.1.4   Ad Hocness and Free Parameters

In Chapter 3 we saw that both Worrall (2014) and Hitchcock and Sober (2004), building on Forster and Sober (1994), suggest that the ad hocness of a hypothesis is related to the free parameters in a theory. More specifically, for Worrall, a hypothesis H is ad hoc with regard to some evidence E when it contains free parameters that need to be fixed on the basis of E for H to accommodate E. As we saw in Chapter 3, Worrall's rationale of why it is bad for a theory to possess free parameters seems Popperian: a theory takes no risk of being refuted by evidence that can be accommodated by fixing its parameters. But we have already criticized the Popperian rationale appropriately. For Forster and Sober, in particular, a hypothesis is ad hoc if it overfits the data (and accommodates noise in the data) with an excess number of free parameters and therefore does poorly with regard to future data. We also saw that both accounts seem to lack some plausibility when compared against Mendeleev's periodic table of chemical elements. Forster, Sober, and Hitchcock's conclusions apply to data models; they do not extrapolate to higher-level theories.

5.1.5   Subjectivist Accounts of Ad Hocness

In contrast to the two accounts discussed in the previous two sections, there are also approaches that deny that ad hocness can be explicated in any particular way. Such approaches tend to treat ad hocness as subjective projections or aesthetic judgements.

In a remarkable paper discussing the role of the Michelson–Morley experiment, the historian Holton noted that scientists' judgements about a hypothesis being ad hoc are often accompanied by the hypothesis in

---

[15] Worrall (2014) points out that the LFC should not be viewed as a conjunct to Lorentz's ether theory, since a theory entailing a falsity (such as Lorentz's ether theory prior to LFC) will still entail a falsity when a conjunct is added to it (for any conjunct).

question being characterized in aesthetic terms as 'artificial', 'contrived', 'strange', 'surprising', and the like (Holton 1973, 327).[16] Holton also claims that ad hoc judgements are highly context-dependent and vary inter- and even intra-subjectively. That is, what might be regarded as ad hoc by some may be seen as non–ad hoc by others, and what might be at regarded as ad hoc at one point in time, according to these authors, might be regarded as methodologically sound at a later point in time, even by the same person (Holton 1973, 176–183). In conclusion, Holton urged that philosophical analysis 'must be supplemented by an understanding of matters of scientific taste and feeling' (183). Recently, Hunt (2012) affirms that 'scientists' aesthetic sense or "feeling" governs their judgments in this matter [of ad hocness]' and that the answer to the question of whether a hypothesis is ad hoc is 'largely in the eye of the beholder' (13). In Section 5.3.2, we will discuss some of the historical evidence that has been cited in support of the subjectivist account. Suffice it to say here, though, that the inter-subjective and intra-subjective variance of ad hoc judgements, if it is real, does not entail the conclusion that the subjectivists try to establish. Scientists make mistakes. Thus, when scientists disagree as to whether a hypothesis is ad hoc or not, this does not have to mean that each scientist has legitimate reasons for their judgement. On the contrary, some scientists might simply be wrong in cases of disagreement. Scientists might also change their minds about the ad hoc status of a hypothesis for *all the wrong reasons*.

I do not want to claim that any variance in ad hoc judgements is illusionary. What I will argue, though, is that such variance can be accommodated within an objectivist account. A general objection against the subjectivist account is this: if ad hoc judgements are purely subjective aesthetic judgements, then whether or not a hypothesis is deemed ad hoc (by someone) ought to have no bearing whatsoever on the confirmation of the hypothesis in question. Indeed, this is the conclusion Hunt draws: 'At the end of the day there seem to be no *ad hoc* hypotheses and no non–*ad hoc* hypotheses, only hypotheses – full stop' (Hunt 2012, 13). But this clearly flies in the face of scientific practice. Scientists use aesthetic language not just to express their 'feelings' and 'tastes'. Ad hoc judgements are normative judgements that imply that one ought not to construct hypotheses in this way and that hypotheses that are so constructed deserve less confirmation than ones that are not ad hoc. But it is not just the scientists'

---

[16] Holton even claims that an ad hoc judgement need not be pejorative and that there are 'acceptable ad hoc hypotheses'. Holton provides no evidence for this claim other than *stating* that scientists sometimes describe them positively (Holton 1973, 327).

judgements that speak against a subjectivist understanding of ad hocness. If ad hoc hypotheses would be just as good as any other non–ad hoc hypotheses, then there would be very little constraint on theorizing. Any theory facing anomalies could be amended at will without that being objectionable (as pointed out already by Popper; see Chapter 1).

5.1.6 Extant Accounts Concluded

Let us take stock. Ad hocness as the lack of independent testability has been defended only by a few philosophers (if any) after Popper's own example of the LFC was shown to be independently testable indeed. The independent support account of ad hocness, despite its popularity amongst philosophers, does not accord with the historical facts either. The Neptune hypothesis, a standard example used by its proponents, was never considered ad hoc. There are no indications that independent support for the LFC would have made it more acceptable, and the discovery of the Higgs particle does not eliminate any of the features of the Higgs mechanism that have been deemed ad hoc by physicists. Parameter accounts of ad hocness (which follow from their novel success counterparts) lack workable rationales, as we have pointed out already in Chapter 3. Subjectivist accounts of ad hocness are built on a non sequitur: the variance in ad hoc judgements alone does not entail their subjectiveness.

5.2 A Coherentist Conception of Ad Hocness

In this section, I will now finally present my own account of ad hocness. Let us start out with another example. The Ptolemaic theory of our planetary system famously used the theoretical device of epicycles: the planets were envisaged to move in circles whose centre in turn moved along the circumference of other circles around the central body, believed to be earth. Although epicycles have become proverbial for ad hoc devices, it was not those to which Copernicus objected when criticizing the Ptolemaic system in his *De revolutionibus*. In fact, Copernicus himself made heavy use of them (Kuhn 1957).[17] Instead, what appears to have taken central stage in Copernicus's objections to the Ptolemaic system were concerns about coherence:

---

[17] Gingerich (1975) raises doubt that the numbers of epicycles can be determined unequivocally in any of the two systems.

[Ptolemaic astronomers have not] been able ... to discern or deduce the principal thing – namely the shape of the universe and the unchangeable symmetry of its parts. With them it is as though an artist were to gather the hands, feet, head and other members for his images from diverse models, each part excellently drawn, but not related to a single body, and since they in no way match each other, the result would be a monster rather than a man. (Copernicus, cited in Kuhn 1957, 137–8)

I will further specify what underlies Copernicus's criticism in a moment, but I take it to be prima facie evidence for what I call the Coherentist Conception of Ad Hocness (CCAH), according to which a hypothesis H is ad hoc, iff theory T and H do not cohere *or* H does not cohere with the accepted background theories B.[18] More precisely:

> **Ad hocness:** A hypothesis H, when introduced to save a theory T from empirical refutation by data E, is ad hoc, iff (i) E is evidence for H and (ii) H appears *arbitrary* in that H coheres neither with theory T nor with background theories B – i.e., neither T nor B provides *good reason* to believe that H (possibly specifying a particular value of a variable) rather than non-H (or some value other than the one specified by H).

Condition (i) is trivially true for any ad hoc hypothesis, for ad hoc hypotheses get introduced to accommodate the data, which T cannot, in the first place. Condition (ii) prominently features coherence relations. Coherence has been described as a measure for 'how well things hang together' (BonJour 1985). There is no agreement about what exactly coherence amounts to, but I will assume that if H coheres with T or B, T or B will give one *good reasons to believe* that H rather than non-H. These reasons, by virtue of being provided by T or B, will be *theoretical* reasons. How are theoretical reasons to believe H to be understood? And when are theoretical reasons good ones?

For T or B to provide theoretical reasons to believe that H (rather than non-H), T or B must of course be logically consistent with H. But consistency is not sufficient for T or B to provide theoretical reasons to believe that H (rather than non-H), just as consistency is not sufficient for coherence.[19]

---

[18] McMullin (1998, 133–4) was perhaps the first to link ad hocness with the lack of coherence. Yet McMullin fails to spell out systematically what that might mean. He appears to construe coherence in terms of causal-unificationist explanation. In his discussion of the Copernican system in astronomy, for example, he remarks about Copernicus that 'he is saying that a theory that makes causal sense of a whole series of features of the planetary motions is more likely to be true than one that leaves these features unexplained' (134). See the next section for my reservations about the unificationist account.

[19] In the following, I will, for the sake of simplicity, not write out the contrastive part of 'reasons to believe H *rather than non-H*'. Whenever I do use the simplified phrase, I nevertheless want it to be

For T or B to provide theoretical reasons to believe that H, further conditions must be satisfied. In the most straightforward case, T or B provide theoretical reasons to believe that H when H can be deduced from T or B. But there are other ways of theoretically justifying H. For example, a theoretical reason to believe that H might consist in an explanation for *why* H should be the case, by the lights of T or B. Such an explanation, in turn, could be causal, or it could consist of a subsumption of H under a regularity. T or B might also provide reasons to believe that H simply in virtue of ruling out possible scenarios inconsistent with H (as we shall see in a moment). In other words, I believe it is advisable to adopt a pluralist attitude towards the kinds of permissible theoretical reasons to believe that H, which T or B can provide, just as one might want to be pluralist about the nature of coherence relations. Lastly, we will demand that for theoretical reasons provided by T or B to be *good* ones, T or B will have to have *empirical* support. Otherwise, any arbitrarily cooked-up theory could be used to render an ad hoc hypothesis non–ad hoc. At the same time, it should be emphasized that, by the lights of CCAH, good reasons to believe that H can never be *just* empirical reasons alone, for they must be provided by T or B, and thereby be theoretical reasons. Let us now illustrate CCAH in more detail with the Ptolemy–Copernicus example.

Copernicus explicated the aforementioned analogy by discussing a number of observations which in the Ptolemaic system had to be simply assumed but for which the Copernican system gave good reasons to believe.[20] Some of these observations were the maximum elongation of the inferior planets, i.e., the observational fact that Mars and Venus are never observed beyond a certain angle from the apparent trajectory of the sun on the celestial sphere, the so-called ecliptic (28° and 47°, respectively). In order to account for this fact, it was decreed in the Ptolemaic system that the centre of the epicycle on which an inner planet would move had to be fixed on a line connecting the sun and the earth. In the Copernican system, in contrast, the inner planets cannot possibly move away from the sun beyond a certain angle, because the inner planets' orbits are encompassed by the earth's orbit around the sun (see Figure 5.1).[21]

---

understood in its complete version. The contrastive part ensures that the reasons provided by T or B are *relevant* reasons to believe that H.

[20] Another point that was important to Copernicus was that the Ptolemaists had, with the deployment of eccentrics, departed from the Aristotelian principle of uniform motion. For a highly interesting discussion, see Miller (2014).

[21] For a detailed discussion, see also Janssen (2002a).

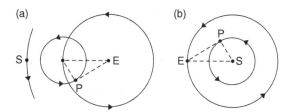

Figure 5.1 Maximum elongation of the inner planets. Ptolemaic system (a) and the Copernican system (b).
From Kuhn 1957. With permission of Harvard University Press.

The way in which the Ptolemaic system accommodates the phenomenon (E) of maximum elongation is clearly ad hoc, because neither the Ptolemaic system itself, nor the background theories at the time provided any good reasons to believe the hypothesis H that the centre of the inner planets' epicycles must remain fixed on a line connecting the sun and the earth. H was arbitrary since theoretically unmotivated. In contrast, in the Copernican system it is *impossible* for the inner planets to move beyond a certain angle away from the sun (as observed from the earth), as it was surmised (H*) that the inner planets' orbits are located inside earth's. The Copernican system thus provides excellent theoretical reasons to believe that H*, as H* follows straightforwardly from the Copernican system.

Before illustrating CCAH with further examples, let us clarify CCAH and compare it with other accounts of ad hocness.

5.2.1 Clarifications and Comparisons with Other Accounts

Let us first note that, contrary to a not-too-uncommon misconception, saving a theory from refutation by invoking a hypothesis is not sufficient for that hypothesis to be ad hoc. Consider again the stipulation of Neptune after the detection of irregularities in the orbit of Uranus (cf. Section 5.1.1). Again, contrary to common folklore, there is no evidence that the hypothesis of Neptune was considered ad hoc before Neptune was discovered, even though it was introduced to save Newtonian mechanics. On the CCAH, the Neptune hypothesis was never ad hoc, because Newtonian mechanics *did* provide good reasons to believe that there was another planet in the vicinity of Uranus, given the irregularities of its orbit.

Second, I would like to point out that my coherentist account of ad hocness is an objectivist account. When a hypothesis does not cohere with either T or B, H will be ad hoc. Contrapositively, when H coheres with T or B, it won't be ad hoc. Yet, the account does allow for a subjective element in ad hoc judgements when it comes to weighting the relevant relations: some scientists might judge some connections of H to T or B to be more important than others. I'm therefore proposing an analogue to Kuhn's model of theory choice where individuals weigh objective theory properties according to their own (subjective) preferences (Kuhn 1977a; see Chapter 2). Analogously, scientists can legitimately differ about which coherence relations they consider to be more important (in Section 5.3.2, we shall discuss an example of a historical dispute where two camps of scientists evaluated coherence relationships differently). But that has no bearing on whether or not coherence relations obtain between H and T or B. Accordingly, judgements about some hypothesis H being ad hoc when as a matter of actual fact H does cohere with T or B are *false* judgements. Subjectivist accounts of ad hocness (Section 5.1.5) have no such restrictions. My account thus sets clear limits to the legitimacy of ad hoc judgements.

Relatedly, proponents of a subjectivist account of ad hocness have cited degrees of ad hocness as evidence for ad hoc judgements being entirely aesthetic. Yet again, one can admit the former without drawing the latter (radical) conclusion. CCAH, too, allows for *degrees* of ad hocness: the stronger the coherence between H and T or B, the less ad hoc is H. The strength of coherence may perhaps be gauged both in terms of the number of relations and the quality of these. It is also worth noticing that degrees of ad hocness allow us to say that one theory ought to be preferred over another, regarding ad hocness, even if neither theory manages entirely without ad hoc hypotheses: one theory might just invoke fewer ad hoc hypotheses than the other. In fact, I think one might question whether there is any prominent scientific theory which does entirely without any ad hoc assumptions.

Take again the Copernican system: it did not manage to do entirely without ad hoc assumptions. For example, it is a consequence of the Copernican system that one should be able to observe a parallax shift, namely an apparent shift in the position of stars, due to the observer's changing positions on earth moving around the sun. Yet the effect could not be observed up until the early nineteenth century because of technological limitations (telescopes were not advanced enough at the time). Copernicus thus had to invoke the idea that the stars were in fact much farther away from us than generally thought at the time. However, this hypothesis was incoherent with the accepted background theories of the time and therefore ad hoc. This was just one of the reasons why it took decades before the Copernican system gained wider acceptance (Copernicus also lacked a workable physics for a moving earth). Regardless, the Copernican system was much less ad hoc, as compared with the Ptolemaic system, when it came to accounting for observations regarding the planets, the sun, and the moon. For example, it explained in a non–ad hoc fashion the fact that only the planets, but not the sun and the moon, retrogress and why the frequency of planetary retrogressions decreases from Saturn, Jupiter, and Mars but increases from Venus to Mercury, and also that the superior planets are the brightest in their opposition (Copernicus 1543/1992 26–7). Thus, we can say that the Copernican system, when it was first proposed, constituted an improvement with regard to the number of ad hoc assumptions it had to invoke in accounting for many planetary observations, which made it preferable over the Ptolemaic system. At the same time, the Copernican system, too, had to make do with a number of ad hoc assumptions before it was eventually developed into a system coherent with accepted physics. Thus, the CCAH account, although allowing the basis of comparative theory choice on the basis of degrees of ad hocness, does not suggest simplified algorithms that do not adequately mirror the complexity of scientific practice.

Of course, there are other important considerations in theory choice than ad hocness. Most importantly, a theory needs to be empirically adequate. And if we do not find a theory that is empirically adequate *and* non–ad hoc, clearly we should stick to it in any case if it is empirically adequate. I think this is true of the Higgs mechanism after the Higgs discovery, which we discussed in Section 5.1.2. Similarly, Planck in 1900 famously introduced the concept of quantization of energy into physics in an ad hoc fashion in order to be able to derive the frequency distribution and temperature dependence of black-body radiation. Planck assumed that light could be emitted and absorbed only in amounts of energy equal to the product of the frequencies of light and integral multiples of what is now known as the Planck constant (Kuhn 1987). Planck's hypothesis is a paradigmatic ad hoc hypothesis: there were no reasons for introducing it *other than* to derive his black-body radiation law. But of course everything panned out rather nicely eventually, and Planck today is celebrated as sparking a scientific revolution. Thus, although theoretically undesirable, ad hoc hypotheses can drive progress not only by way of helping scientists save the phenomena, but also by giving them incentives to develop theories which render those very same hypotheses non–ad hoc (by providing good theoretical reasons for believing in them). Hence, the view of methodological norms that recommends itself is not one where norms are categorical, but rather conditional or ceteris paribus. That is, the norm not to devise ad hoc hypotheses has to be weighed against other norms such as 'seek hypotheses that are empirically adequate'. It can therefore not always be wrong to devise ad hoc hypotheses; it is admissible when no other hypotheses are available. It is nevertheless always desirable to devise hypotheses that are not ad hoc, as the aforementioned example illustrates.

One may also wonder about how CCAH relates to the other accounts of ad hocness that I mentioned earlier, in particular the parameter-fixing account and the account that construes ad hocness in terms of lacking unification. Let us start with the former. On the parameter-fixing account of ad hocness, fixing free parameters on the basis of the data, so as to render the theory in question empirically adequate, counts as ad hoc. CCAH is compatible with the parameter-fixing account: when a theory has free parameters and those are fixed on the basis of the data the theory is supposed to accommodate, then this is ad hoc by the lights of CCAH, because there are no good *theoretical* reasons for fixing the parameters at values that accommodate the data in question. We can thus accept that parameter-fixing is undesirable but reject the rationales provided by their proponents (see Section 5.1.4 and Chapter 3).

Because the fixing of free parameters is also discouraged in CCAH, CCAH accommodates the fact that physicists have deemed the Higgs mechanism ad hoc for (i) its large number of free parameters (cf. Section 5.1). The other three grounds for ad hoc charges frequently mentioned by physicists are that (ii) there are no other fundamental scalar particles other than the Higgs boson, (iii) the symmetry-breaking of the Higgs mechanism is different from all other known cases of symmetry-breaking, and (iv) the Higgs particle is 'fine-tuned' to its interaction mass and unexplained. Whereas the last point, too, falls under the heading of 'T provides no reason for believing that H', points (ii) and (iii) are instances where B provides no reason for believing that H, as our background theories give us no reason why there ought to be only one fundamental scalar particle and why symmetry-breaking may proceed differently only in the Higgs mechanism.

In Section 5.1.2, we also mentioned the discovery of the Higgs boson and the Higgs mechanism (which requires the former) as one example amongst others that seems to undermine the independent support notion of ad hocness. One may nevertheless wonder whether one could perhaps hold CCAH in conjunction with the independent support condition, in case the independent support condition would get its plausibility from other examples. Let us briefly ponder this possibility.

On the independent support view of ad hocness, it is sufficient that a hypothesis H have independent support for it not to be ad hoc (likewise, it is necessary that a hypothesis have independent support for it not to be ad hoc). Yet, on my account, what is needed for H not to be ad hoc is that there be theoretical reasons for belief in H. Independent empirical support is thus not sufficient for non–ad hocness, on my account. In fact, it's not even necessary: even when a hypothesis does not have independent empirical support (but only the evidence it was invoked for), it could count as not ad hoc on my account (namely, exactly when there are good theoretical reasons for belief in H). So my coherentist conception and the independent support notion appear to be incompatible.

Let us now consider the relation of CCAH to another account of ad hocness. Coherence is intuitively closely related to unifiedness. But although several writers have gestured at a lack of unifiedness when discussing ad hoc features of theories (cf. Section 5.1.3), a precise account of what this would amount to is as of yet wanting. Suppose the rough idea is that an explanation is unified if it explains a set of phenomena with a small amount of basic assumptions or principles, and say that we possessed a particular unified explanation/theory of a certain set of phenomena (Kitcher 1981). Would that mean that *none* of the phenomena in that set could be accommodated in an ad hoc fashion by the relevant theory? Or would it mean only that at least *some* of the phenomena should not be accommodated in an ad hoc fashion? The former is implausibly restrictive: many highly unifying theories make ad hoc assumptions (see, e.g., the standard model and our discussion of the Higgs mechanism in this chapter). But if it's only some phenomena, what is it in the unification account that determines which phenomena are not ad hoc with regard to a unifying theory? In other words, the unification account of ad hocness has a *specificity* problem which is absent on the CCAH. In fact, unification and CCAH are nicely complementary: a unified theory can be coherent to one or another degree, and, vice versa, a coherent theory can be unified or not (or maybe unified to one or the other degree).[22] At the same time, a unified theory will never lack *potential* coherence, as it presumably will give reasons for believing in at least some of the lower-level hypotheses about the phenomena which it unifies (for actual coherence to obtain, those reasons must of course be good reasons).

[22] It has in fact been argued that unification accounts subscribe to a winner-take-all conception of explanation according to which an explanation either is or is not explanatory, without allowing for degrees of explanation (Woodward 2014a).

There is another problem of the unification account, which I alluded to already in Section 5.1.3 and which the coherence account avoids. As mentioned, unification is usually understood as capturing a number of phenomena with relatively few assumptions. Unification accounts of ad hocness view the need to invoke ad hoc hypotheses for a theory to accommodate the relevant evidence (roughly) as a sign of the lack of unifiedness of a theory. But this idea is not unproblematic. Suppose we have two sets of phenomena P_1 and P_2 and two theories T_1 and T_2, both of which manage to accommodate P_1 and P_2, albeit in different ways. Whereas T_1 accounts for both P_1 and P_2 in terms of its own basic assumptions, T_2 accommodates P_1 but must invoke additional hypotheses in order to be able accommodate P_2. Intuitively, we would want to prefer T_1 over T_2. Does the unificationist account provide the right resources for justifying this preference? First of all, note that T_1 and T_2 accommodate the same sets of phenomena. Whether T_1 or T_2 would count as unifying by the lights of the unificationist account therefore hinges on the number of principles invoked. Yet, determining this number and whether or not T_1 or T_2 is superior in that regard is by no means a trivial task: as we know from Duhem, both T_2 and T_1 will have to invoke a multitude of auxiliary assumptions in order to be able to accommodate the phenomena, all of which will have to figure in the final count of assumptions. We cannot therefore know in advance whether T_2 or (the intuitively preferable) T_1 would come out as more unifying. But this indeterminacy is not tolerable on an account that seeks to provide a principled way of analysing ad hocness. The proponent of the unificationist account may be able to address this problem by giving us a principled way of distinguishing between fundamental and less fundamental assumptions in a theory, but before this is achieved, the coherence account must seem superior to the unificationist account of ad hocness.[23] To this we may add that the unification account, contrary to CCAH, cannot accommodate the fact that background knowledge seems to play a role in ad hoc judgements, as in the case of the Higgs mechanism (see earlier) and as we shall see in another example in detail later, in Section 5.3.2.

In Chapter 3 we noted that accounts of novel success are motivated by a desire to guard against ad hocness. So, one may ask, how does the CCAH account relate to predictive success, and in particular the thesis that predictive success is better evidence for a theory than accommodative success? On CCAH, there is no asymmetry between evidence that is

---

[23] In fact, Worrall has struggled with this question and concludes that there is no good way of doing so (Worrall 1989b, 2002, 2005).

predicted and evidence that is accommodated. In both cases, CCAH requires good reasons for the hypothesis invoked to predict or accommodate E. So, for example, when Einstein, in 1915, was able to derive from the theory of general relativity (T) that light would bend around massive objects such as stars (H), the theory provided reasons for believing in the existence of starlight-bending, which motivated Eddington and others to go out and collect observational evidence (E) for H. In contrast, the Ptolemaic system, which we briefly considered previously, can for example predict that the inner planets will not be observed beyond a certain angle from the sun, but it gives no good reasons for believing in this prediction (because its stipulation of coordination of the position of the sun, the earth, and the centre of an inner planet's epicycle cohere neither with the theory itself nor with the background knowledge). Thus, although the Ptolemaic system is able to make predictions regarding maximum elongation, the basis on which it does this is objectionable. Predictive success therefore does not secure against ad hocness.[24]

Finally, one might perhaps be worried that the CCAH could commit one to a coherentist epistemology for science, with all its well-known and highly problematic implications. But that is not the case. Coherentism is the view that a belief is justified *if and only if* it coheres with other beliefs. Nothing in the CCAH would imply that much. On the contrary, I believe that theories and hypotheses in science are justified when they are supported by the relevant evidence, just as foundationalism has it. Nevertheless, those support relations, on CCAH, will be stronger or weaker depending on the amount of coherence between the hypothesis and the theory in question. In other words, the coherence relations *modify* the support relations. The epistemology to be associated with the CCAH may therefore be referred to as weak foundationalism (Olsson 2012).

5.3 In-Depth Illustrations of the Coherentist Conception

In this section I want to discuss two of the aforementioned historical cases in further detail. In particular, I want to argue (i) that an attempt by

---

[24] The notion of ad hocness defended here is in some ways similar to the one hinted at by Lange (2001), discussed in Section 5.1.3 and Chapter 3. Once more, however, I would want to lay claim on having presented a more general and systematic account (cf. fn. 18). And some points made by Lange I simply don't agree with. For example, Lange alleges that 'a *few* examined cases do not suffice to lead scientists to formulate [an arbitrary conjunction] ... Therefore, a hypothesis judged to form an arbitrary conjunction typically arrives with *many* accommodations' (577; emphasis added). I don't see at all why this ought to be so in general.

Lorentz to render the Lorentz–FitzGerald contraction hypothesis non–ad hoc by providing theoretical reasons to believe it supports the CCAH and (ii) that the cosmological controversy between proponents of the steady-state theory and the Big Bang theory show that disagreements about the respective ad hoc status of those theories can be understood in terms of subjective weightings of objective coherence relations.

5.3.1   The Lorentz–FitzGerald Contraction Hypothesis Revisited

In the analysis of the LFC, one must distinguish between an early and a later, mature version. The early version was proposed in cursory form by Lorentz in 1895 in a short section in the final chapter of a book 139 pages (and by FitzGerald around the same time) and the mature version in 1904. Many philosophers of science, for various reasons, believe that the ad hoc charge applies only to the early, but not to the mature, version of the LFC (Zahar 1973a; Grünbaum 1976; Janssen 2002b; Acuña 2014). That is consistent with us possessing evidence for scientists such as Einstein deeming the early version of LFC ad hoc, while there is no information about what Einstein might have thought about the later version of LFC (Holton 1969, 169).

Let us briefly recall that, in the early version of LFC, Lorentz simply assumed that (i) 'molecular forces are also transmitted through the ether, like the electric and magnetic forces of which we are able at the present time to make this assertion definitely'; that (ii) the 'attraction and repulsion' of molecular forces for a body at rest would be in equilibrium; and that (iii) the Lorentz transformations would apply not only to electrostatic forces, but also to the 'molecular forces' holding together matter (Lorentz 1875, in Einstein et al. 1952, 6). As Lorentz readily admitted himself, 'there is no reason' in his theory for making the latter assumption in particular. And assumptions (i) and (ii) are also highly implausible (cf. Janssen 2002b, 437).

Some commentators have emphasized that Lorentz's more mature and general theory from 1899 also was able to account for second-order effects in very precise versions of the Michelson–Morley experiment (Schaffner 1974; Janssen 2002b, 425).[25] It also entailed *further* testable consequences, such as the velocity dependence of mass (Janssen 2002b). But even with the

---

[25] For Janssen (2002b, 425), the 'generalized' LFC is the following: 'a matter configuration producing a certain field configuration in a frame at rest in the ether will, when the system is set in motion, change into the matter configuration producing the corresponding state of that field configuration in the frame moving with the system'.

mature LFC, it has been argued, there is something wrong. According to Janssen (2002b),

> [i]n Lorentz's theory, there is a strict separation of ether and matter ... Lorentz decreed a number of important exceptions to the Galilean-invariant Newtonian laws that are supposed to govern matter, so that the laws effectively governing matter are Lorentz invariant. Why, one can legitimately ask, would the laws governing matter have the property of Lorentz invariance, which so far appeared to be nothing but a peculiar property of Maxwell's equations? ... In the final analysis, it is thus left an *unexplained coincidence* in Lorentz's theory that both matter and fields are governed by laws that are Lorentz invariant [whereas in Einstein's theory of special relativity, it isn't]. (423 and 426; emphasis added)

For confirmation of this assessment, Janssen (2002a) cites Poincaré's dismissal of the LFC in his introduction of *Sur la dynamique de l'electron* (1906):

> We cannot content ourselves with simply juxtaposing formulas that would agree only by some happy coincidence; the formulas should, so to say, penetrate each other.

And indeed, Lorentz was merely *assuming* also in the mature theory that the Lorentz-transformations would apply to 'molecular forces', holding together matter, in the same way they applied to electrostatic forces (Lorentz 1904, in Einstein et al. 1952, 22).

Janssen does not link this perceived deficit of Lorentz's theory to judgements about LFC being ad hoc. Indeed, he concludes that 'a solid case can be made for the claim that [Lorentz's mature] theory is not ad hoc by any of the criteria considered here' (437). Instead, he regards the unexplained coincidence in Lorentz's mature theory as a *different* reason for why Lorentz's theory was inferior to Einstein's.[26] Yet I don't see why one ought to keep those reasons separate. After all, already in the publication of the early LFC version, as we mentioned before, Lorentz admitted that 'there is no reason' to suppose (as he did) that the Lorentz transformations should apply also to matter. But if that's so, then it's not too implausible to suppose that it is *this* that Einstein and others objected to when deeming the (early) LFC ad hoc, and what they would have objected to also in Lorentz's mature theory.

It is interesting to note that Lorentz, in a letter to Einstein in 1915 – i.e., 10 years after proposing his mature theory – stated his belief that the LFC

---

[26] See Acuña (2014) for a detailed criticism of Janssen's account.

was rendered non–ad hoc by him offering an explanation for it in terms of molecular forces:

> I had added that one can arrive at this hypothesis [i.e., the LFC], if one extrapolates from what one was able to say about the influence of translation on electrostatic forces to other forces. *Had I stressed it more, the hypothesis would have made less of an impression of having been devised ad hoc.* (Lorentz 1915 in Schulmann et al. 1998, 71–2)[27]

What Lorentz had said in 1875, again, was

> that [the LFC] is by no means far-fetched, as soon as we assume that *molecular forces are also transmitted through the ether*, like electric and magnetic forces of which we are able at present time to make this assertion definitely ... *From the theoretical side*, therefore, there would be no objection to this hypothesis. (Lorentz, in Einstein et al. 1952, 6; emphasis added)

So what Lorentz appears to have thought is that the LFC lost its ad hoc character at the moment when he was able to lend to it some *theoretical* plausibility. And this he thought despite the fact that he at the same time admitted that he had devised the molecular forces explanation only *after* he had come up with the LFC (Lorentz 1915, in Schulmann et al. 1998, 74). Although we don't know what Einstein made of that suggestion,[28] Lorentz doesn't seem to be alone in this judgement. Leplin (1975, 314–5, fn. 18) points out, for example, that two later textbooks (one from 1924 and one from 1969) seem to suggest that 'Lorentz's representation of contraction as a condition of molecular equilibrium mitigated its *ad hoc* character'.[29]

From the point of view of CCAH, what Lorentz appeared to have sought to do in order to diminish the ad hoc status of the LFC was to establish a coherence relation between the LFC and the rest of the ether theory.[30] That

---

[27] This is my own translation of the original German text.

[28] In his reply to Lorentz, Einstein did not mention the issue (Schulmann et al. 1998).

[29] Zahar (1973a) suggests that it is for this reason that the LFC is not to be regarded ad hoc. Zahar claims that Lorentz was able to 'derive' the LFC from the molecular force hypothesis. But that's not the case. Lorentz offered only a 'plausibility' argument, not a derivation (Janssen 2002b, 436–7). In his mature theory, Lorentz derived the length contraction from what Janssen calls the *generalized* LFC (see fn. 25).

[30] One might be tempted to interpret Lorentz's molecular forces hypothesis as an attempt to produce an explanation that would engender novel predictions and that it was for the latter reason, not the former, that Lorentz thought the LFC was rendered non–ad hoc. But, as we have already noted in Section 5.1.1 with Holton, independent predictions of the LFC 'were not urged as tests that would decide on its acceptability' (Holton 1969, 177). It should also be noted, once more, that Lorentz's early (non-generalized) LFC was already testable with the Kennedy–Thorndike experiment (Janssen 2002b, 433).

he achieved only to a limited degree. Although there perhaps was *some* plausibility in assuming that the molecular forces that Lorentz postulated for matter would behave not unlike the electromagnetic forces 'since both types of force are states of the same substratum', as Zahar (1973a, 116) put it, it remained highly curious how this was to be achieved. As mentioned previously, the 'molecular forces' hypothesis required that there be an electrostatic equilibrium when a body is at rest. But there is no such thing as electrostatic equilibrium (cf. Janssen 2002b). Thus, although Lorentz was able to provide *some* reasons for belief in LFC, he wasn't able to provide *good* reasons. His attempted explanation of the LFC in terms of molecular forces did not establish coherence with either the ether theory or with the background theories.

5.3.2   Extended Case Study: The Cosmological Controversy

As we've seen in Section 5.1.5, some writers have claimed that ad hoc judgements are *nothing but* aesthetic judgements. The main supporting example of the most recent contribution in this school of thought (Hunt 2012) is a selection of the cosmological controversy between proponents of the Big Bang theory and its main competitor, the steady-state theory, in the 1950s–1960s. This controversy is rich and telling. In particular, I believe that it provides a nice illustration of both the idea that ad hoc judgements are the result of subjective weightings of objective coherence relations and the idea that background knowledge should be part of the equation in the analysis of ad hoc judgements. [31]

Both the Big Bang theory and steady-state theory presume that the universe is expanding, and (in agreement with Hubble's law) that the galaxies farthest away from us are the fastest. [32] In the forward time direction, expansion will result in a decrease in the density of matter and energy. In the words of Hoyle (1955), 'space is therefore (it seems) getting more and more empty as time goes on' (315). In the backward time direction, density increases and the universe 'contracts'. It would seem that an *origin of the universe back in time* should have been a natural conclusion, since, going back in time far

[31] The following significantly expands on Hunt's (rather cursory) discussion by drawing substantially on Kragh (1996) and some original sources.

[32] E. Hubble in 1929 provided the first observational data for an expanding universe, although Hubble himself did not interpret his results in this way. Before it became widely accepted that the universe is expanding, Einstein had inserted the notorious cosmological constant into his general theory of relativity in order to keep the universe static. After the Hubble discovery, Einstein removed the constant and called his earlier appeal to it his 'greatest blunder'. See Kragh (1996). Without Einstein's constant, an expanding universe follows naturally from Einstein's theory.

enough, all matter of the universe should be 'squeezed into a uniform mass of very high density' (Gamow 1961, 28). Yet the idea of an origin of a universe wasn't so natural to many physicists at the time. In particular, the proponents of the steady-state theory sought to evade this consequence. They subscribed to the so-called *perfect* cosmological principle, which had it that the distribution of matter and energy must appear 'the same' to an observer anywhere in the universe at any time. As we shall see in a moment, steady-state theorists embraced this principle, because they believed that it was required by the idea that laws of physics applied throughout the universe (cf., e.g., Bondi 1960/1952).

Although steady-state theorists, too, accepted that the universe is expanding, they believed that the universe, essentially, has always looked the way it does today. They viewed expansion not as resulting from a Big Bang, but rather from a dynamic equilibrium of attractive and repulsive gravitational (Newtonian) forces, at short and large distances, respectively. In order to theoretically counterbalance a decrease of the density of matter in space over time as implied by expansion, steady-state theorists postulated the continual creation of matter, evenly distributed throughout the universe. Indeed, in order to maintain the steady-state picture of the universe, one had to assume that matter is being continually created in such a way that 'the effect of expansion is [*precisely*] compensated in such a way that the total amount of matter in the observable universe remains constant' (Hoyle 1949, 18). And it was only for this reason that one could deduce the rate of creation from the mean density of matter (which was thought to be constant) and the rate of expansion (as given by Hubble's law); it was estimated to be three hydrogen atoms per cubic metre per million years (Kragh 1996, 183). However, and crucially, there appeared to be no grounds for supposing that the rate of creation of matter ought to coincide *precisely* with the expansion rate *other than* to save the perfect cosmological principle.

From the beginning, the continual creation hypothesis was viewed with great suspicion by the majority of astronomers and physicists. For one thing, steady-state cosmologists were accused of introducing an unnecessary additional assumption in the form of the continual creation hypothesis (McVittie 1949, 49; Milne 1949; cf. Kragh 1996, 190). For another, that self-same hypothesis was inconsistent with the conservation of energy, as implied by Einstein's theory of relativity. Furthermore, and perhaps most importantly, whereas the Big Bang theory could simply employ Einstein's field equations of general relativity, the proponents of the steady-state alternative struggled badly to come up with a workable dynamics. Some provided a merely qualitative theoretical framework (Bondi and Gold 1948). Others, who tried to devise a modified field-theoretic treatment (Hoyle 1948), faced the objection of self-defeat with regard to its goal of providing a unique solution, as compared to the flexibility of the evolutionary relativistic models, such as the Big Bang theory, as we shall see in a moment (cf. Kragh 1996, 205). Later formulations faced even more serious obstacles (Kragh 1996, 213). But even if the steady-state theorists had managed to come up with a workable mechanics, this would have left at least some of the critics unimpressed, for

> [if the steady-state theory is] accepted, it is necessary to suppose that the mechanics of the nebular systems are in some way *different from the mechanics of all other astronomical systems*. If the object of science is to unify phenomena into theoretical systems with as wide an amplitude as possible, the general relativity may be accepted until it leads to some prediction seriously contrary to observation. (McVittie 1951, 75; emphasis added)

Let us now follow the thread of Hunt's story in his case for his subjectivist view of ad hoc judgements (Hunt 2012). Hunt focuses on the cosmologist D. W. Sciama (1926–1999). He claims that Sciama adopted the steady-state hypothesis and not the Big Bang theory because he considered the latter, but not the former, ad hoc. According to Hunt, Sciama disfavoured the Big Bang theory in particular for its parameter flexibility and the arbitrary setting of an initial temperature for the creation of heavy elements (Hunt 2012, 7). In 1960 Sciama, for example, wrote that the 'actual behavior of the universe can be accounted for [with the steady-state hypothesis] without *ad hoc* assumptions' (Sciama 1960, 10, cited in Hunt 2012, fn. on p. 11), implying that the Big Bang theory couldn't accomplish such a feat.

Continual creation, for Sciama, wasn't as problematic as it was for many other physicists. On the contrary, Sciama thought that '[i]f creation [of matter] is occurring all the time [by virtue of the continual creation hypothesis], it becomes a scientific process you can study *because it's repetitive*'. The Big Bang theory, in contrast, Sciama considered 'an awkward thing' (both preceding quotations from an interview with George Gale on 27 January 1990, cited in Hunt 2012, 7). Other proponents of the steady-state theory, such as Bondi and Gold, also repeatedly emphasized this point (Kragh 1996, 181ff.). For example, Bondi and Gold (1948) could 'see no reason why the laws of nature should be invariant while admitting that the one and only application [i.e., to the universe] is not invariant [but rather a singular event]'.

Hoyle, another major proponent of the steady-state hypothesis, wrote that '[a]n explosive creation of the Universe [as envisaged by the Big Bang theory] is not subject to analysis. *It is something that must be impressed by way of an arbitrary fiat*' (Hoyle 1955, 318; emphasis added). It was also for this reason that Hoyle considered the Big Bang theory ad hoc.[33] 'In the case of a continuous origin of matter, on the other hand', Hoyle continued, 'the creation must obey a definite law, a law that has just the same sort of logical status as the laws of gravitation, nuclear physics, of electricity and magnetism' (Hoyle 1955). Yet what this law was supposed to look like the proponents of the steady-state theory were never able to specify. And as mentioned earlier, no reason was given why matter creation should proceed at *exactly* the rate that would counterbalance the extension of the universe *other* than that this would make a steady universe possible *despite* expansion. Indeed, Bondi, admitted that '[t]he theory offers no explanation of the[se] numerical coincidences' (Bondi 1960/1952, 151). So the accusation of arbitrary fiat applied to them as much as to their opponents, if it did at all.

The continual creation hypothesis inspired the search for the stellar (rather than cosmological) creation of elements, and it also was able to sport some success explaining it (Kragh 1996, 295). However, Gamow was not impressed at all with the explanations that it offered. Echoing Copernicus's aforementioned misgivings about the lack of coherence in the Ptolemaic system, he wrote that what the theory effectively required in terms of processes of element creation was similar to

> the request of an inexperienced housewife who wanted three electric ovens for cooking a dinner: one for the turkey, one for the potatoes, and one for the pie. Such an assumption of heterogeneous cooking conditions, adjusted to give the correct amounts of light, medium-weight, and heavy elements, would completely destroy the simple picture of atom-making by introducing a complicated array of specially designed 'cooking facilities'. (Gamow 1961)

Furthermore, Hoyle et al.'s theoretical treatment had a decisive defect: it could not account for the creation of the amount of the light element

---

[33] With regard to the formation of galaxies, Hoyle stated explicitly: 'In the explosion theory the formation of clusters of galaxies has to be introduced as an *ad hoc* process that takes place for no good reason at just the stage where the density of matter falls to a thousand million million million millionth part of the density of water (or perhaps somewhat less than this)'. In contrast, the steady-state hypothesis, Hoyle thought, offered 'a more natural explanation' (Hoyle 1955, 317).

helium, one of the two most abundant elements in the universe (together with hydrogen) (Kragh 1996, 338).

One important criticism that steady-state proponents levelled against the Big Bang theory, in turn, had to do with parameter freedom. Bondi and Gold (1948), in their foundational paper, wrote that

> [i]n general relativity a very wide range of [cosmological] models is available and the comparisons [between theory and observation] merely attempt to find which of these models fits the facts best. The number of free parameters is so much larger than the number of observational points that a fit certainly exists and not even all the parameters can be fixed. (262)

Kragh (1999) has accordingly referred to relativistic cosmology as 'not a theory in the ordinary sense [but] rather a supermarket of theories which had in common that they were all solutions of the same fundamental [Einstein field] equations' (378). In a similar vein, Sciama (1961) wrote that

> [a] theory which, whilst it can be tailored to fit this unique universe, nevertheless *has to present certain aspects of it as arbitrary, as though they could have been different*, is therefore less satisfactory than a theory in which these aspects are essential. (7; emphasis added)

In the same paper, Sciama presented the steady-state theory as a theory 'in which the *actual* behavior of the universe can be accounted for without *ad hoc* assumptions' (10; original emphasis). In his book, Sciama expanded on this thought, writing that

> [w]hat the cosmologist requires, therefore, is a theory which is able to account in detail for the contents of the universe. To do this completely, it should imply that the universe contains no accidental features whatsoever. This provides us with a criterion for assessing the validity of rival theories. We believe this criterion to be so compelling, that the theory of the universe which best conforms to it is almost certain to be right. (Sciama 1959, 150; cf. Sciama 1960, 323)

Sciama was confident to have found such a theory in the steady-state theory. Again, Sciama contrasted the steady-state theory with the Big Bang alternative. In particular, he criticized that on the Big Bang theory the size of galaxies was used to determine the character of the early universe – in particular, temperature fluctuations. Sciama complained that 'no reason is given' why the early universe should have one particular fluctuation characteristic rather than any other, *other than* such a characteristic providing the initial conditions for deriving the current form of our galaxies. Such an assumption, Sciama criticized, was merely 'accidental' and 'devoid of theoretical significance' (Sciama 1959, 150).[34] Sciama had a similar criticism of the Big Bang's explanation of the origin of heavy elements:

> But to my mind there is a more important reason for preferring the steady-state theory. For in theories which start from an explosion [i.e., Big Bang–type theories] *the initial properties of the universe are entirely arbitrary.* Thus it is possible to find an initial temperature which is favourable for making heavy elements, and then one *simply has to assume* that this was the initial temperature although *the general theory would be equally compatible with any other initial temperature.* This means that in this type of theory the laws of physics do not specify the contents of the universe, but only show how one state of the universe follows from another. (Sciama 1955, 42; emphasis added)

Up until the 1960s, there was no evidence that could tell in favour of either the Big Bang or the steady-state theory (Kragh 1996, 269ff.). Once it finally became available, it worked against the steady-state theory: abundant quasars and radio galaxies were found at large distances, but not in closer galaxies. According to the steady-state theory, such inhomogeneities were puzzling. The proponents of the steady-state theory, however, tried to save their theory despite these difficulties. Sciama (again, Hunt's main protagonist in his case study concerning ad hocness) sought to explain the abundance of radio sources at greater distances by locating our part of the galaxy in a 'local hole' of nearby galactic radio sources (rendering the observed abundance only an *apparent* one). He also demanded more data before ruling out local quasars and proposed an astronomical mechanism *mimicking* the black-body nature of the microwave background. At some point (around 1966), however, Sciama gave up the steady-state hypothesis and in fact criticized those who put even more 'epicycles' on it to account for the forthcoming observations.[35] Hunt concludes that 'scientific judgments about "how much was too much" were quite different' (Hunt 2012, 9) and that '[a]s scientists' aesthetic sense or "feeling" governs their judgments in this matter [of ad hocness], this will manifest itself in different ways' (12).

5.3.2.1   Assessment
Let us now assess this episode from the point of view of CCAH. Recall that one of the main criticisms of the steady-state theory (SST) made by the Big

---

[34] Interestingly, the formation of such a structure in the history of the universe is an unsolved problem until this very day.

[35] He for instance refrained from extending the local 'hole' in the distribution of radio sources and insisted that it was 'impossible that any quasars could be at cosmological distances' (Hunt 2012, 8).

Bang theory (BBT) proponents appears to have been that SST was inconsistent with both the general theory of relativity and conservation principles. The continual creation of matter hypothesis (CCM), which the steady-state theorists postulated in order to save the phenomena, did not cohere with the relevant background theories and was therefore ad hoc. Another criticism – which was perhaps not levelled as explicitly[36] – was that SST gave *no reason for belief* in CCM and, in particular, no reason for the assumption that the rate of creation of matter ought to coincide *precisely* with the expansion rate of the universe (as admitted explicitly by Bondi; cf. Section 5.3.2). The only reason that the proponents of SST could provide for CCM was that the perfect cosmological principle was sustainable together with the known expansion rate *only* under the assumption that CCM would compensate precisely for the expansion rate. We are thus very much reminded of Lorentz's attempt to save his ether theory by stipulating length contraction: there was no reason in his theory to believe that 'molecular forces' would obey the letter of the Lorentz transformations just as the electrostatic forces did. The only *sustainable* reason he could provide was that it saved the appearances (the molecular forces hypothesis, as we saw in Section 5.3.2, was an attempt to give a theoretical reason, but a failed one). Both the contraction hypothesis and the CCM are ad hoc by the lights of CCAH because neither T nor B provided good (theoretical) reasons for believing them. That is, in neither case did T or B provide good reasons for believing that the Lorentz transformations ought to extend to matter, and, in the case of CCM that matter ought to be created at a rate that would compensate for the expansion of the universe.

The SST proponents, on the other hand, criticized the BBT on two counts. First, they objected to BBT postulating a beginning of the universe, i.e., a unique, non-repetitive event which for that very reason would be different from other events subject to the laws of nature. Hoyle, in particular, went as far as saying that the Big Bang would *exempt* the beginning of the universe from the laws of nature and considered this highly arbitrary and ad hoc. Furthermore, SST theorists were discontent with the fact that there was a multitude of possible cosmological models resulting from the fixing of parameters in the relativistic field equations that underlay the BBT. For this reason, the BBT can thus also be

---

[36] I was not able to find any explicit use of the word 'ad hoc' by Big Bang proponents. Yet the normative remarks by Milne and McVittie, mentioned earlier, I take to be (non-explicit) ad hoc judgements. Also, Gamow commented on several occasions that the continual creation hypothesis was 'unnecessary' (Gamow 1954, 60; 1961, 34) and 'artificial' (Gamow 1952, 40). As Holton has pointed out, scientists use such terms interchangeably with ad hocness (see Section 5.1.4).

considered ad hoc (to some limited extent) on the coherentist conception of ad hocness: it gave no reason for belief in fixing the free parameters of the relativistic field equations to any particular values in BBT theorists' attempts to account for structure and element formation in the early universe (see Section 5.3.2).

In sum, we can thus say that were objective grounds for *both* sides of the debate to accuse the other camp of making ad hoc assumptions – namely, the lack of coherence relations. The reason practitioners disagreed about the merits of the two theories is explained by the different weightings of the coherence relations (or the lack thereof) by different scientists. That is, we can accommodate a subjective element of ad hoc judgements on CCAH without rendering ad hoc judgements entirely arbitrary, as the subjectivists would have it. Likewise, CCAH can accommodate the re-evaluations of ad hoc judgements by individual scientists, as in the case of Sciama: over time, coherence relations may be established between T or B and H where there were none before.

In the final assessment, I think it is fair to say that the assumptions made by the SST theorists were more problematic than the ones in BBT. Again, in contrast to BBT, the SST was inconsistent with two background theories (the conservation principle and the general theory of relativity) *and* could provide no reason for belief in the creation rate coinciding precisely with the expansion rate. It is thus understandable, from the coherentist conception of ad hocness, why the broad class of evolutionary relativistic models, which BBT was an instance of, was able to draw much more support from the community of physicists and astronomers than the steady-state theory even *before* decisive empirical information could be garnered.

5.3.3   The Bohr Model and the Kinetic Theory and Specific Heat Anomaly Revisited

It is worth noting that CCAH also nicely accounts for the changes made to the Bohr model of the atom and the kinetic theory in response to various spectral data and the specific heat anomaly, respectively, as discussed in Chapter 4. In the case of the Bohr–Sommerfeld model, background theories, and in particular the theoretical apparatus developed for our solar system, gave physicists good reasons to believe that electrons in the atom would analogously follow elliptical orbits, rather than just circular orbits. Contrary to McMullin, we need not appeal to any physical necessity to de-idealize to make sense of (at least some of) the changes to the Bohr model being non–ad hoc. Nor should we, as it is seems problematic in general to speak of physical necessity in the Bohr–Sommerfeld model, where several seemingly established physical principles had to be given up.

In the case of the kinetic theory of heat and the specific heat anomaly, similar observations apply. When Boltzmann introduced his dumb-bell model, he did not have any good theoretical reasons (either by virtue of the kinetic theory or by virtue of background theories) to believe that diatomic molecules behave like dumb-bells. This is why his model was judged ad hoc. Yet, later such reasons *were* provided by quantum mechanics, which is why the model ceased to be ad hoc.

5.4 The Third Virtuous Argument for Realism: The Argument from Coherence

As we saw in Chapter 2 in our discussion of pessimistic meta-induction (PMI), realists have been keen to stress the generation of novel success as a necessary condition for their commitment to a theory's approximate truth. This move not only significantly shortens Laudan's list of false but successful theories, but it also provides a guide for the divide et impera strategy, namely the identification of truth-candidates amongst parts of our rejected theories. In Chapter 3, however, we argued that there does not seem to be a robust rationale for novel success being somehow intrinsically superior to non-novel empirical success. The realist's response against the PMI thus appears unsupported. And yet, as we shall see in a moment, realism can be defended without novel success being superior evidence.

What many of the (unconvincing) proposals privileging novel success share is the intuition that novel success is better evidence because it guards against ad hoc moves: if the evidence E is not known at a time a hypothesis H entailing E is introduced, H cannot be ad hoc with regard to E; if it is not allowed that E be used in the construction of H, then H cannot be ad hoc with regard to E; etc. McMullin's account of fertility, too, is driven by the intuition that ad hoc hypotheses should be avoided (Chapter 4). Instead of seeking means to guard against ad hocness, I proposed in the current chapter that we might as well shift the focus of the discussion to a *direct* analysis of ad hocness.

On my coherentist account of ad hocness (CCAH), there is no privilege for novel success either. So long as the hypotheses introduced to accommodate anomalies cohere with the theories they amend (or with the background theories), the evidence in question is confirmatory. A lack in coherence results in a decrease in confirmation. With this notion of ad hocness in hand, we note a potential advantage for the realist: since coherence, on my account, is to be understood in terms of *theoretical* reasons for belief, and since ad hocness clearly affects theory confirmation, theoretical reasons for belief affect theory confirmation. Unless the anti-realist denies that ad hocness affects confirmation, which would very strongly run counter to scientific practice and very common intuitions about theory construction, and unless she can show that CCAH is inadequate, she is thus forced to admit that theoretical reasons for belief affect theory confirmation. This, in turn, puts the antirealist in a rather uncomfortable position: she would have to accept that reasons for belief stretch beyond empirical reasons for belief. However, this is a view which she strongly rejects. If she would accept it, she would also see her arguments weakened for the underdetermination of theories by evidence: given two empirically adequate theories, we would have good reasons to believe in the theory that is less ad hoc/more coherent.

With regard to the PMI, the realist might now try to argue that the past and false theories on Laudan's list were in fact not as coherent as more recent theories. For example, the Ptolemaic astronomical system, as we saw earlier, was not as coherent as the Copernican system. And the Big Bang theory was more coherent with important background theories than the steady-state theory. Both the Copernican system and the Big Bang theory have turned out to be much closer to the truth than those competitors. This is no coincidence. When a theory is coherent in my sense, it provides good theoretical reasons for belief in those of its parts which account for the phenomena. The more coherent a theory is, the fewer the number of hypotheses that need to be invoked in order to account for the phenomena for which the theory (or background theory) provides no reasons for belief. But what is it that makes some theories more coherent than others?

Consider a set of phenomena caused by the same structure. For this set of phenomena, clearly, a coherent explanation is available – namely, the one which refers to that structure. A theory referring to this structure could then provide good reasons for belief in those of its parts which account for the phenomena. Conversely, if there is no common structure for the phenomena the theory accommodates, then there is no discoverable explanation connecting the phenomena. The reasons for belief which a theory could provide for any of its parts accounting for the phenomena wouldn't be good reasons. Thus, in principle, a theory's coherence is a sign for it correctly identifying a real structure underlying the phenomena that it accounts for.

Of course, we may be mistaken about *thinking* that we possess good reasons for belief in parts of a theory, and accordingly mistaken about the phenomena being connected (these would be the false positives of discovery). Likewise, we may be wrong about not possessing good reasons for belief when such reasons are in fact available and the phenomena are indeed connected (these would be the false negatives). Yet, in a well-functioning science, we should expect that scientists are more likely to find coherent theories when there *are* such structures (and the true positive rate to be higher than the false positive rate; cf. Chapter 2). Thus, in well-functioning science, theoretical coherence (in my sense) should increase the probability of a theory being true. This is my *third virtuous argument for realism* or my *argument from coherence*.

Of course, the theories that we will ever be able to devise are unlikely to be absolutely true. Likewise, I believe, our theories are unlikely ever to be fully coherent, or free of ad hoc accommodations. There will always be hypotheses we need to invoke which lack good theoretical reason for belief and which cannot be grounded in the identified structure, simply because our knowledge of reality will always be limited.

CHAPTER 6

Virtues as Confidence Boosters and the Argument from Choice

In Chapters 1, 2, and 5, I presented my first three *virtuous arguments for realism* – namely the *argument from simplicity*, the *no-virtue-coincidence argument*, and the *argument from coherence*, respectively. In this chapter, I provide my final virtuous argument for realism – the *argument from choice*.

The argument from choice seeks to undermine the view that theoretical virtues are merely pragmatic and not at all epistemic, by taking a close look at the theory-choice decisions made by scientists in some groundbreaking discoveries. The view in question I want to call the Negative View about theoretical virtues.

Here is how this chapter proceeds. In Section 6.1, I set out the Negative View and what I call the dictatorship condition, which follows from it. In Section 6.2, I present historical cases that undermine the dictatorship condition and, by implication, the Negative View. In Section 6.3, I spell out my *argument from choice* and then consider objections in Section 6.4.

6.1 The Negative View, the Dictatorship Condition, and Its Violation

Realists contend that scientists pursue true theories not only on the basis of a theory accommodating the facts, but also on the basis of a theory's unifying power and simplicity. Antirealists like van Fraassen (1980, 1989) and others (e.g., Douglas 2009),[1] however, deny that those virtues are epistemic. For them, particularly the virtue of simplicity is merely pragmatic: it only characterizes the convenient *use* of a theory. A mathematically simple theory, for example, will be easier to comprehend and handle than one that is complex. This is why theoretical virtues have also been described as features of a theory that merely 'make our minds feel

---

[1] Cf. fn. 33 and 35 in Chapter 1.

good' (Hacking 1982). The only virtues the antirealist considers epistemic are a theory's empirical strength, its internal and external consistency, and, most importantly, its empirical adequacy – i.e., a theory's observable consequences of the past, present, and future being true. The Negative View which antirealists embrace is that the other theoretical virtues are *not at all epistemic* – i.e., that theoretical virtues such as simplicity and unifying power have got nothing to do with a theory's truth.

Empirical adequacy, like truth, is an ideal limit which we might never achieve. All we can say at a certain moment in time is that our best theories have been empirically adequate *until now*. We cannot know now whether our theories will be empirically adequate in the future; our theories might have consequences which we haven't yet been able to test, evidence which we thought confirmed a theory might turn out false, etc. All we can do, therefore, is *aim* for empirical adequacy. Antirealists grant that theoretical virtues, which they regard as merely pragmatic, may influence our decision to commit to one or the other theory in the pursuit of this goal. We may commit to a theory on the basis of its simplicity, for example, and hope that our commitment to it will in the future be vindicated by the theory's empirical consequences turning out to be correct. Van Fraassen, however, insists that commitment has nothing to do with the epistemic content of a theory: 'Belief that a theory is true, or that it is empirically adequate, does not imply and is not implied by, belief that full acceptance of the theory will be vindicated' (van Fraassen 1980, 13). Van Fraassen here is appealing to the inductive risk we are taking when believing a theory. The world might cease to exist tomorrow. In that case, the theories we decided to embrace will not be vindicated. Yet that wouldn't make false our belief that our current best theories are empirically adequate (to the extent that they are). Thus, given the inductive risk that is involved in the acceptance of a theory, *belief* in the empirical adequacy of a theory (which, as we just saw for van Fraassen, is the only belief implied by the acceptance of a theory) reaches only as far as a theory's empirical adequacy *up until now*.

Even though van Fraassen holds that the pursuit of theories (on the basis of their virtues) has nothing to do with justified belief, he still thinks it is *rational* for us to engage in such pursuits, because the search for virtuous theories cannot be separated from the search for empirically adequate theories (88). Indeed, according to van Fraassen, there are no explanatory (i.e., pragmatically virtuous) theories that are not empirically adequate. As he puts it, 'having a good explanation *consists* for the most part in having a theory with those other qualities [namely empirical adequacy]' (94), and 'we don't say we have an explanation unless we have an *acceptable* theory which explains' (95). This is why the pursuit of explanatory theories is fully compatible with the empiricist view. For instance, the decision to pursue the Copernican theory of the planets was rational despite it being empirically on a par with the Ptolemaic system in the sixteenth century, because it was fully consistent with what the empiricist considers the aim of science, namely empirical adequacy. The pursuit of theories with pragmatic virtues, van Fraassen concludes, might therefore be 'the best means to serve the central aims of science', i.e., empirical adequacy (89). But what if the pursuit of explanation and empirical adequacy come apart? Could it ever be rational for a scientist to adopt a virtuous theory inconsistent with the facts? Here van Fraassen is adamant:

> This can surely be not so when the other virtue is the one the non-realist regards as the highest – empirical adequacy. For to forgo empirical adequacy is to allow that there may arise inconsistencies with observed facts. That possibility we cannot allow while advocating a theory as correct. (van Fraassen 1980, 94–5)

In other words, a theory may never be adopted and 'advocated as correct' – i.e., believed to be empirically adequate – if it is not in fact empirically adequate, regardless of how otherwise virtuous the theory might be. This is of course very much in tune with the Negative View. Conversely, believing a virtuous but not empirically adequate theory, on van Fraassen's view, is irrational.

Let us refer to the condition that a theory may not be believed when it is not empirically adequate, regardless of how virtuous it is, as the *dictatorship condition*.[2] I want to argue that the dictatorship condition, in a number of important historical cases of theory choice, has been violated. The cases I have in mind are the following:

- Einstein's general theory of relativity and early data on light-bending
- Watson and Crick's model of DNA structure and antihelical evidence
- Mendeleev's periodic table and data of already-known chemical elements
- Einstein's special relativity and the early Kaufmann's experiments
- The Vine–Matthews–Morley hypothesis (a precursor of plate tectonics) and early data on sea-floor magnetization

---

[2] The term is inspired by an important, much discussed contribution to the problem of theory choice (Okasha 2011). In the context of Okasha's discussion, a dictatorship condition would imply that a theory T_1 which is ranked above another theory T_2 with regard to some 'dictator' virtue X would *overall* be ranked above T_2, regardless of how T_1 and T_2 rank with regard to other virtues.

- The Glashow–Weinberg–Salam model (a precursor of the standard model in fundamental particle physics) and early data on the weak neutral current

In all of these cases, scientists sought to assess the theories in question (T). This was complicated by the fact that the evidence available at the time was conflicting: there was evidence for and against T. In all of these cases, the evidence against T was eventually dismissed as unreliable by proponents of T – despite the fact that they had little evidence for the unreliability of the evidence contradicting T. The evidence for T they treated as confirmation of T. Proponents of T did not just pursue T but believed it; otherwise, they would not have dismissed as unreliable evidence contradicting T and would have withheld judgement as to whether or not T was confirmed or disconfirmed. There are further cases of similar importance (cf. Schindler 2013b), but in what follows I will restrict myself to a discussion of the aforementioned ones.

Obviously, these cases are all instances of theoretical bias. Not all kinds of theoretical bias are malign, though. Suppose a theory predicts p_1, p_2, and p_3. We carry out experiments whose results come out in support of p_1 and p_2. Now we conduct experiments regarding p_3, and these experiments produce evidence contradicting p_3. On the basis of our inductive support from experiments regarding p_1 and p_2, it would not be unreasonable to then harbour doubts about the negative results regarding p_3 and to perhaps conduct further tests. Think of the report from a few years ago concerning the OPERA experiments that neutrinos travel faster than the speed of light, in blatant contradiction of a central postulate of Einstein's theory of special relativity (Reich 2011). The immediate response of the physics community was disbelief and suspicion that some experimental error would account for the result.[3] Given the previous tremendous empirical success of Einstein's theory, it would have been simply foolish to firmly focus one's disbelief and suspicion on Einstein's theory rather than the data. Even before the experimental error was actually found (caused by an incorrectly connected fibre-optic cable in the timing system), such a reaction was indeed reasonable (Reich 2012).[4]

Einstein's theory of special relativity is a very well confirmed theory. The cases we will now consider are not of this kind. Instead, the theories in question were newly proposed and had little inductive support. In order to

[3] Cf., e.g., Overbye (2011).
[4] I have called this use of theory in the reliability assessment of data as *theory-driven data reliability judgements* (Schindler 2013b).

avoid the conclusion that scientists acted utterly irrationally with regard to advancing science (and committed grave methodological mistakes), we must accept that the scientists in question adopted these theories for good epistemic reasons. They would have had such reasons if the Negative View were false and theoretical virtues were indeed epistemic. Theoretical virtues would then legitimately have boosted scientists' confidence that the theories in question were correct – despite the fact that some of the evidence suggested their falsity. This, in essence, is my *argument from choice*, which I will develop further in Section 6.3. But let us first attend to the empirical support for my argument.

6.2 Historical Case Studies

6.2.1 *Einstein's General Theory of Relativity and Early Data on Light-Bending*

As already mentioned earlier in this book, Einstein's general theory of relativity is often considered to have been impressively confirmed in 1919 by evidence for its (temporally novel) prediction of starlight-bending, gathered by British astronomers led by Eddington. In a remarkable paper which re-assessed the evidence for light-bending in 1919, however, Earman and Glymour (1980) concluded, contrary to folklore, that 'the British results, taken at face value, were conflicting and could be held to confirm Einstein's [general] theory *only if many of the measurements were ignored*' (Earman and Glymour 1980, 51). Let us review some of the details.

The basic technique employed to measure starlight-bending by Eddington and his colleagues consisted in superimposing photographic plates, taken with telescopes, of the observed positions of stars of a particular star field during a solar eclipse and during the night (see Figure 6.1).

Although seemingly straightforward enough, this basic procedure was complicated by various confounding effects. First and foremost, astronomers had to make sure that the apparent changes of star positions did not result from a 'change of scale' between the eclipse plate and the comparison plate, so that a millimetre on each plate would correspond to different seconds of arc (the measurement unit used back then for starlight-bending). Such changes of scale could be caused by mechanical changes in the telescopes due to, for example, temperature changes, which were to be expected given that the photographic plates were taken months apart (because one had to wait for the sun to move out of the star field).

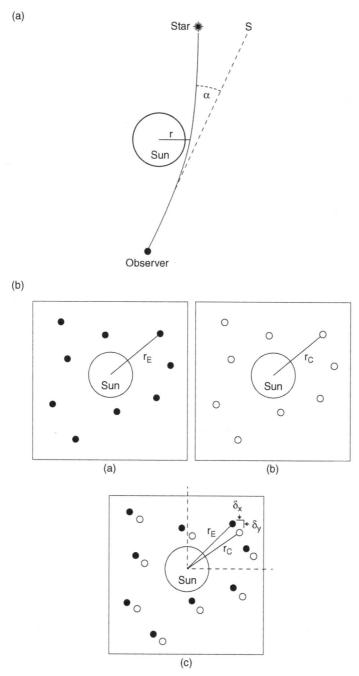

Figure 6.1 Starlight-bending and its measurements. Left: the phenomenon of starlight-bending causes stars to appear in positions that differ from their actual ones. Right: (a) eclipse plate (with the sun present), (b) comparison plate (during the night), and (c) superposition of eclipse and comparison plate with measurements of apparent position changes.
Adopted from von Klüber (1960) with permission of Elsevier.

Another error source was position, due to changes in the orientation of the plates to the optical axis.

The eclipse expeditions led by Eddington produced three data sets from two expeditions (one to Sobral in Brazil, one to the island of Principe off the West African coast), which can be labelled according to the kinds of telescopes which were used: (1) the Sobral 4-inch, (2) the Sobral astrographic, and (3) the Principe astrographic. Whereas data set 1 (1.98″ ± 0.178″) and data set 3 (1.61″ ± 0.444″) were roughly consistent with the prediction of Einstein's theory (1.75″), data set 2 (0.93″ ± 0.48) was inconsistent with Einstein's theory but in agreement with the prediction of Newton's theory (0.87″). Let us, accordingly, refer to the data sets 1, 2, and 3 as Sobral(+), Principe, and Sobral(−), respectively.[5]

In their report published in the *Philosophical Transactions of the Royal Society* in April 1920, the British astronomers discounted Sobral(−) on the basis of a *possible* change of scale (Dyson et al. 1920)[6] and did not even mention it at their prior presentation of their results at the 'Joint eclipse meeting of the Royal Society and the Royal Astronomical Society' on 6 November 1919 (Thomson 1919). Earman and Glymour have objected to this discounting of Sobral(−) for the following reason. Since Principe was 'the worst of all', Earman and Glymour argue, 'it is hard to see decisive grounds for dismissing one set [i.e., Sobral(−)] but not the other [i.e., Principe]' (Earman and Glymour 1980, 74–5). Although Earman and Glymour deem Sobral(+) 'much more impressive' than Sobral(−) and Principe, they remark with respect to Einstein's theory that 'the mean value [of Sobral(+)] is too high and the dispersion too small' (74–5). Effectively, Earman and Glymour point out, Sobral(+) was no better evidence for Einstein's theory than Principe was for Newton's theory.[7] Earman and Glymour infer that physicists presupposed a 'trichotomy of possible results' that they never argued for: the results would either indicate no starlight deflection, a deflection consistent with Newton's theory, or

---

[5] I here quote the standard deviations provided by Earman and Glymour (1980). The original article by Dyson et al. contained 'probable errors'. It is also noteworthy that the raw data of Sobral(−) indicated a light deflection of 0.86″ rather than 0.93″. The latter value was gained only after restricting the analysis to a few particularly bright stars. See Dyson et al. (1920).

[6] In their paper, Dyson et al. (1920) wrote that 'the images [of Sobral(−) were diffused and apparently out of focus ... These changes must be attributed to the effect of the sun's heat on the mirror, *but it is difficult to say whether this caused a real change of scale in the resulting photographs or merely blurred the images*' (309; added emphasis). In other words, it was uncertain (by the author's own lights!) whether the images were ruined by confounders or not.

[7] As can be gleaned from Dyson et al.'s data, Sobral(+) is about 1.3 standard deviations away from the Einsteinian value and Principe about 1.7 standard deviations from the Newtonian value.

a deflection consistent with Einstein's theory. Only then, and under the supposition of Sobral(–) being somehow erroneous, 'the results had to be viewed as confirmation of Einstein's prediction' (79–80). Earman and Glymour reach a similarly gloomy verdict on Einstein's redshift prediction. They conclude: 'If one were willing to throw out most of the data, one could argue that Einstein's prediction was confirmed' (51).

It is pretty obvious that some kind of theoretical bias in favour of Einstein's general theory of relativity is at play in the analysis of Eddington and his colleagues.[8] As we saw in Section 6.1, though, theoretical bias is not necessarily bad; sometimes there are very good reasons for it, resulting in reasonable questioning of the relevant data. Again, it would have been anything but foolish not to be sceptical towards the alleged discovery that neutrinos travel faster than the speed of light. Physicists were justified in their scepticism because, at that point, special relativity was just a very well confirmed theory. Crucially, of course, the general theory of relativity in 1919 was not yet well confirmed. Indeed, the confirmed prediction of light-bending usually gets cited as the first and most impressive confirmation of Einstein's theory. And yet it's not correct that Einstein's theory had no confirmation whatsoever before the light-bending test – unless, that is, one is not willing to confer *any* confirmational power whatsoever to phenomena known before a theory enters the stage. In Chapter 3 we noted that this view, under whichever interpretation of novelty, is not plausible. We also learned in the same chapter that Einstein's successful (non–ad hoc) explanation of Mercury was received very positively in the physics community, and even more so than the confirmation of the light-bending prediction (Brush 1989). Indeed, when presenting the eclipse results to the Royal Society and the Royal Astronomical Society in 1919, Sir Dyson, Astronomer Royal and member of the eclipse expeditions, stressed that Einstein's theory 'had already explained the movement of the perihelion of Mercury – long an outstanding problem for dynamical astronomy' (Thomson 1919). Eddington seconded that Einstein's theory was confirmed by Mercury's perihelion (393). The president of the Royal Society, Sir Thomson, admitted that 'it is difficult for the audience to weigh fully the meaning of the figures that have been put before us' but expressed trust in the interpretation suggested by Eddington and his collaborators and concluded that Einstein's theory 'has

---

[8] Kennefick (2007) has presented three arguments against Earman and Glymour's assessment which I have argued to be ineffective (Schindler 2013b).

survived two *very severe tests* in connection with the perihelion of Mercury and the present eclipse' (Thomson 1919; emphasis added).

There is no *direct* evidence that Eddington or his collaborators, or the community of astronomers who received their results, were motivated by Einstein's success with Mercury to view the eclipse results in a light favourable to Einstein's theory. But unless one is willing to invoke socio-logical causes (Collins and Pinch 1998), this seems to be the most plausible interpretation.

6.2.2 Watson and Crick's Model of DNA Structure and Antihelical Evidence

In 1953 Crick and Watson discovered that the structure of DNA is a double helix consisting of a phosphate backbone and A-T and G-C base pairings. Rosalind Franklin produced evidence for the structure of the DNA by means of x-ray crystallography, some of which Crick and Watson used – not uncontroversially – in building their metal and cardboard models of the DNA.

In 1951–1952 Franklin realized that DNA, depending on its water content, would produce two distinct x-ray diffraction pictures. She called the dry form the A-form and the wet form the B-form. Although the A-form would yield sharper spots on the diffraction pictures, suitable for crystallographic analysis, the B-form provided some crucial information for Crick and Watson's model building.

Generally, a reciprocal relationship obtains between diffraction pictures and the fibres containing the investigated polymers (which, by convention, are mounted vertically): large spacings in the pattern of the diffraction picture correspond to short spacings in the periodic structure of the polymer, and vice versa. The distances between the layer lines on the photograph are inversely proportional to the *repeat* of the polymer (i.e., a recurring structure such as a helical turn) along the fibre axis. Furthermore, the distances between the spots along the layer lines correspond to repeats perpendicular to the fibre's vertical direction.[9] For a helical structure, that would mean that the distance between the layer lines corresponded to the pitch of the helix, the 'height' of the cross to the rise of the helix, and the angle between the arms of the cross to the width of the helix (see Figure 6.2).

What is sometimes not emphasized enough is that Crick and Watson relied on a helical diffraction theory developed by Crick and two

---

[9] For details, see e.g. Mathews et al. (2000).

(a)   (b)

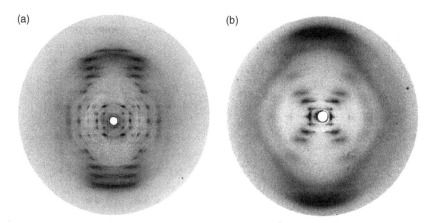

Figure 6.2 Franklin's x-ray diffraction pictures of the A-form and B-form of DNA. Note that the picture of the A-form (left) displays sharp spots, and the picture of the B-form (right) smudges. Both forms produce meridional absences of reflections and strong black arcs on the top and bottom of the pictures, corresponding to the rise of the helix.
With kind permission of King's College London.

collaborators in Cochran et al. (1952) (henceforth, CCV theory). That theory predicted the pattern to be obtained on x-ray pictures from a helical structure (continuous and discontinuous): cross-shaped pattern with meridional absences. This is exactly what the picture of the B-form, produced by Franklin, exhibited (see Figure 6.2). Upon seeing the B-picture for the first time, Watson, in his famous book *The Double Helix*, recalls:

> The instant I saw the picture my mouth fell open and my pulse began to race. The pattern was unbelievably simpler than those obtained previously ('A' form). Moreover, *the black cross of reflections which dominated the picture could arise only from a helical structure*. (Watson 1968, 98; emphasis added)[10]

Spurred on by this picture, and on the basis of information about the chemical constitution of DNA and density measurements provided by Franklin, Crick and Watson pursued their model-building, which basically consisted in seeking to reconcile various stereochemical constraints involving molecular bond angles, distances, and kinds (single or double) within

[10] It was Franklin's colleague Maurice Wilkins at King's College London who showed Watson the B-picture. This has led to much debate about whether Franklin was treated fairly (e.g., Elkin 2003). According to Wilkins (2003), Franklin kept the B-picture, which she had produced in May 1952, in a drawer without doing anything with it, until willingly handing it over to Wilkins before she left King's College in January 1953. It was only then, Wilkins reports, that he showed it to Watson.

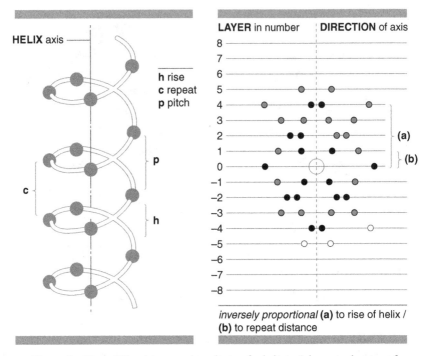

Figure 6.3 X-ray diffraction pattern prediction for helices. Schematic drawing of a helical structure (left) and CCV's x-ray pattern prediction for helices (right). Adopted from Mathews et al. (2000), *Biochemistry*, 3rd ed. Reprinted by permission of Pearson Education, Inc, New York, New York.

a consistent structure.[11] 'The problem' of determining a molecular structure, Crick once reflected, 'is rather like a three-dimensional jigsaw puzzle with curious pieces joined together by rotatable joints [namely, the single bonds]' (Crick 1954, 57).

By several accounts (Wilkins 2003; Klug 2004), Franklin, in contrast, focused all of her efforts on deciphering the structure of the DNA on the basis of the A-form, which was more suitable for applying the so-called Patterson function analysis, essentially a Fourier transform that uses intensities instead of phases.[12] The resulting Patterson maps only represent

---

[11] Crick and Watson started out using false density measurements by W. Astbury, which resulted in a DNA structure with three instead of two strands (L. Pauling made the same mistake). The C2 symmetry group of the DNA was crucial for Crick and Watson's model-building; it implied that the two strands of DNA run anti-parallelly. See Judson (1996, 143 and 102) for details.

[12] Phases cannot be reconstructed from x-ray diffraction pictures. See, e.g., Taylor (1967).

distances between atoms, but not their positions. Franklin did not have much sympathy for model-building and, according to perhaps her closest student, R. Gosling, once said that 'We are not going to speculate, we are going to wait, we are going to let the spots on this photograph [via the Patterson] tell us what the structure is' (Gosling, in Judson 1996, 127; cf. Maddox 2002, 162, 184).

In 1952, the same year in which Franklin discovered the distinct pattern for the B-form, Franklin and Gosling produced a picture of the A-form, which exhibited clear differences in the intensities of the reflections on the two sides of the fibre axis (Olby 1974, 370–1). Franklin took this to be clear evidence against a helical structure. Her colleague Wilkins conceded that he and his colleague Stokes 'could see no way round the conclusion' reached by Franklin: 'It seemed, in spite of all previous indications, that the DNA molecule was lop-sided and not helical' (Wilkins 2003, 182). Although Franklin, by several accounts, accepted that the B-picture was strong evidence for a helical structure, she took her antihelical evidence very seriously, was convinced that some conformational change would occur from the A-form to the B-form, and sought to develop structures that would accommodate it (Klug 1968; Judson 1996, 128). When Franklin confronted Watson with her antihelical evidence, Watson commented – probably with the CCV theory in mind – 'If only she would learn some theory' (Watson 1968, 96). In contrast to Franklin's careful analysis of her antihelical evidence, Crick summarized his and Watson's attitude thus: 'When she told us DNA couldn't be a helix, we said, "Nonsense". And when she said but her measurements showed that it couldn't, we said, "Well, they're wrong"' (Crick, in Judson 1996, 118).

In pondering what made the difference between Crick and Watson's successful discovery of the DNA structure and the failed attempt by Franklin and her colleagues at King's College, Wilkins concludes somewhat hyperbolically:

> *Our main mistake was to pay too much attention to experimental evidence.* Nelson won the battle of Copenhagen by putting his blind eye to the telescope so that he did not see the signal to stop fighting. In the same way, scientists sometimes should use the Nelson Principle and ignore experimental evidence. (Wilkins 2003, 166; emphasis added)

Similarly, Crick quotes Watson as saying: 'no *good* model ever accounted for *all* the facts, since some data was bound to be misleading if not plain wrong. A theory that *did* fit all the data would have been "carpentered" to do this and would thus be open to suspicion' (Crick 1988, 59–60; original emphasis).[13]

The double-helix model by Crick and Watson had several virtues. Most importantly, perhaps, the model reconciled all stereochemical constraints. It was therefore both internally and externally consistent.[14] The model was simple in that it presumed no conformational changes, as Franklin had surmised. It furthermore suggested a simple replication mechanism, as famously noted by Crick and Watson in their discovery paper. It thus promised to be theoretically fertile. These virtues of the model, paired with the CCV's confirmed prediction of the X-shaped x-ray pattern of the B-form, helped to boost the confidence in the helical model proposed by Crick and Watson and contributed towards adopting a critical attitude towards Franklin's antihelical evidence.

6.2.3   *Mendeleev's Periodic Table and Data of Already-Known Chemical Elements*

As we mentioned in Chapter 3, there were several chemical elements which did not fit into Mendeleev's periodic table (most notably, uranium and beryllium). Mendeleev, however, insisted on the correctness of the weight-ordering criterion, which would ensure the coherence of the periodic table, and proposed corrections to the experimental measurements of these elements, which prior to his proposal had been, in Mendeleev's own words, 'generally adopted and seemed to be well established' (Mendeleev 1901). Indeed, Mendeleev's insistence on the correctness of the weight criterion as the all-governing classification criterion, it may be argued, made the difference between Mendeleev's successful classification and previous attempts to classify the elements.

Mendeleev himself distinguished between two types of classification attempts prior to his periodic table: 'artificial' and 'natural' systems. As artificial, he criticized systems which were 'based on some few characters of the elements' (such as affinity, electro-chemical properties, physical properties, their reaction behaviour with oxygen and hydrogen, and their 'atomicity'). As 'natural' he denoted those systems which 'establish, according to members of different and *purely chemical properties*, groups of analogous elements' (Mendeleev 1879, 303; emphasis added). Natural

---

[13]   The Chargaff ratios (1:1 ratios between adenine and thymine, on the one hand, and guanine and cytosine, on the other), contrary to common belief, seemed not to have played any important part in the discovery of the DNA structure. See Judson (1996) and Schindler (2008a).

[14]   Crick and Watson did make a number of basic mistakes on the way, though. See Judson (1996).

systems he in turn criticized for 'not rest[ing] upon invariable principles', for containing elements which 'are without analogies', and for including 'no general expression for the reciprocal relations between the different groups; they could not be united in one whole' (303). Clearly, coherence and unification considerations weighed strongly on Mendeleev's mind – so much so, in fact, that he was prepared to undermine seemingly established experimental results (cf. Chapter 3). We thus witness yet another violation of the dictatorship condition. Mendeleev himself noted that previous attempts to classify the chemical elements were hampered 'because the facts, and not the law, stood foremost in all attempts' and 'wanted [i.e., lacked] the *boldness* necessary to place the whole question at such a height that its reflection on the facts could be clearly seen' (Mendeleev 1889, 638; emphasis added). Mendeleev also pointed out that

> [b]efore the periodic law was formulated the atomic weights of the elements were purely empirical numbers, so that ... [many of its properties] could only be tested by critically examining the methods of determination; in short, we were compelled to move in the dark, *to submit to the facts, instead of being masters of them*. (Mendeleev 1889, 649; emphasis added)

Mendeleev's achievement can be illustrated in comparison with Lothar Meyer's, who is often mentioned as the co-discoverer of the periodic table (as noted earlier on, he in fact received the Davy Medal for the discovery of the periodic table together with Mendeleev). Meyer's first classification attempt, drafted in 1862 and published in 1864 in his book *Die Modernen Theorien der Chemie*, resulted in a main table of 28 elements and a two-part supplementary table of 6 + 16 elements. Only the main table was arranged on the basis of increasing atomic weight and grouped into the valences 4, 3, 2, 1, 1, 2. Scerri (2007) comments:

> This table suggests that Lothar Meyer struggled to arrange elements in terms of atomic weight as well as chemical properties. He seems to have decided to let chemical properties outweigh strict atomic weight ordering in some cases. (96)

In a manuscript prepared for inclusion in the second edition of Meyer's book dating from 1868 (but published only in 1895), Meyer sought to combine the separate tables into one system of now 42 elements. However, that system included aluminium twice, in both cases violating the increasing weight sequence (Smith 1976, 152). Finally, after having read an abstract of Mendeleev's table in the *Zeitschrift für Chemie*, Meyer devised a more complete table, which he conceded 'is essentially identical to that given by Mendeleev'. In contrast to Mendeleev, however, Meyer

(1870) believed that 'the properties of the elements are [only] *largely* periodic functions of the atomic weight', and although he considered atomic weight changes, he concluded that '[i]t would be rash to change the accepted atomic weights on the basis of so uncertain a starting-point [as the periodic system]'. Meyer did not at all comment on the gaps contained in his table. In a subsequent priority dispute between Meyer and Mendeleev, Mendeleev complained that 'Meyer did not comprehend the deeper meaning' of the periodic table, and, accordingly, 'he predicted no atomic weights, and did not alter any' (Mendeleev, cited in van Spronsen 1969).[15]

6.2.4 Einstein's Special Relativity and the Early Kaufmann's Experiments

One of the first tests of special relativity came from a series of experiments on the velocity dependence of the electron by W. Kaufmann between 1901 and 1905. Although Kaufmann's experiments are now known as providing the first novel evidence for Einstein's principle of relativity, they indeed first appeared to provide evidence *against* it. The electron theories by M. Abraham and A. H. Bucherer, which sought to reduce electron mass to the interactions of the electromagnetic field and electron charge, were in much better agreement with Kaufmann's experimental results (Cushing 1981 and Hon 1995).

Kaufmann's experiments were taken very seriously. H. Poincaré, for example, stated in 1908 that they '*have shown Abraham's theory to be right*' (Poincaré 1914, 228; original emphasis). Poincare, despite being very sympathetic to Einstein's theory, could not 'well see what objection can be brought' against the experiments, given that Kaufmann had taken 'all suitable precautions' (229). Likewise, H. Lorentz, who had interpreted Kaufmann's earlier results as vindicating his contraction hypothesis, in fact accepted that Kaufmann's latest data *refuted* his hypothesis (Lorentz 1952, 212–3) and concluded, 'I must abandon it [his theory]. I am therefore, at the end of my Latin' (Lorentz in a letter to Poincare in 1906, cited in Miller 1981, 334). Although he then went on to further explore his theory, regardless, he never considered questioning Kaufmann's results (Hon 1995, 216). In contrast, Einstein, whose theory predicted an experimental outcome hardly distinguishable from Lorentz's, simply ignored Kaufmann's negative results when mentioning his experiments for the first time in print

---

[15] Meyer's table is probably an example of the 'natural systems' criticized by Mendeleev.

and instead proposed a *different* set of experiments to decide between his predictions and the ones of Abraham and Bucherer (see Miller 1981, 333–4). Later, he admitted that Kaufmann's experiments were 'error-free', but speculated about the presence of a systematic error whose 'source he could not determine' (Hon 1995, 209; cf. Einstein 1907). Einstein was furthermore adamant that the theories by Abraham and Bucherer were not correct, despite their better match with Kaufmann's experiments, and he refused to give up on his theory until 'a *greater variety* of observational material will be available' (1907; original emphasis).[16] Not until 1914 would replications of Kaufmann's experiments achieve agreement with Einstein's predictions (Cushing 1981). Given the persuasiveness of the Kaufmann's experiments, and given that Einstein had no evidence to back up his speculation about an experimental error, Einstein's reaction may appear irrational, particularly when compared against, for example, Lorentz's 'incessant attempt to obtain a theoretical framework within *all* experimental results' (Hon 1995, 212; original emphasis). However, it need not be. If Einstein's attitude is not to be accounted for simply by personal bias (as it should not; see Section 6.4), then it is best explained by the virtues of the special theory of relativity (Einstein himself mentioned 'inner perfection' and 'logical simplicity' as important virtues in theory choice; cf. Hon 1995, 209–10). And those virtues can legitimately justify the belief in a theory *despite* contrary evidence, only if they are epistemic.

6.2.5   *The Vine–Matthews–Morley Hypothesis and Early Data on Sea-Floor Magnetization*

In some crucial work in the mid-1960s, which would form the fundamental basis for developing our geophysical paradigm of plate tectonics, Vine and Matthews (1963) and Morley explained some recently discovered, and quite puzzling, magnetizations of the ocean sea floor. The VMM hypothesis, as it is now known, combined the hypothesis of sea-floor spreading (Hess 1962), according to which the sea floor is formed by volcanic activity along mid-oceanic ridges, with the idea of geomagnetic field reversals, according to which the direction of the earth's magnetic field switches in intervals of several millions of years. VMM entailed symmetrical patterns of alternating 'positive' and 'negative' magnetizations parallel to mid-oceanic ridges (see Figure 6.4).

---

[16] This is my translation of the original.

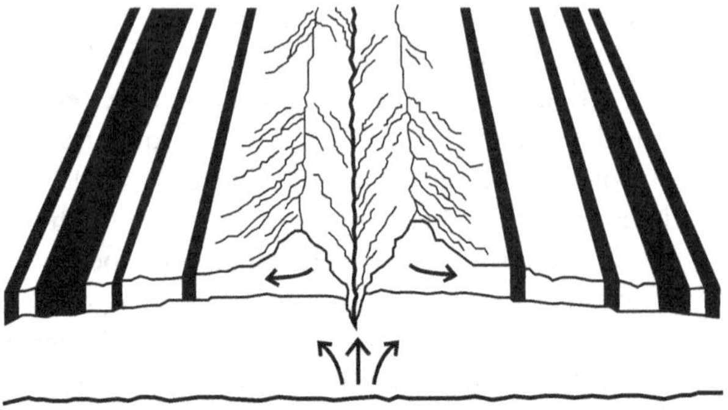

MID-OCEAN RIDGE

Figure 6.4 The Vine–Matthews–Morley hypothesis. Sea-floor spreading and alternating remnant magnetizations of the sea floor (due to geomagnetic field reversals), also known as the 'zebra pattern'.
Figure by Hreinn Gudlaugsson.

Although the VMM hypothesis explained the observed alternating patterns of oceanic magnetization, there were several facts which seemed to contradict this idea. Most importantly perhaps, few magnetic anomaly patterns were as neatly symmetrical as predicted by the VMM hypothesis. It was also found that several patterns exhibited a change from low-amplitude, short wavelength anomalies around the ridge axis to high-amplitude, long wavelength anomalies over the ridge flanks. To some, this difference appeared to be a difference in kind and could not have a common cause, such as the VMM. Accordingly, they sought alternative explanations in terms of the idiosyncrasies of the sea floor (Schindler 2007). Furthermore, the sedimentary and basaltic samples and fossils collected near axes of ridges were too old for sea-floor spreading to be correct (see Frankel 1982). One of the initial sceptics noted with regard to the VMM in 1964: 'this attractive mechanism is probably not adequate to account for all the facts of observation. A theory consistent with the facts is still needed' (Vacquier, cited in Oreskes 2003, 58). A member of the group of geologists at the Lamont Geological Observatory, critical of the VMM, later described their approach as 'data oriented': 'we were people who were out collecting data, and dealt with only theories that could be substantiated heavily with the facts' (Heirtzler, cited in Frankel 1982, 7–38). In contrast, others, such as Vine, defended the VMM *despite* these difficulties. With regard to the change in the amplitude of the anomalies, Vine later remarked, 'I did not attach any significance to such discrepancies . . . I did not consider it relevant to the point I was trying to make' (Frankel 2012, 252). And Hess, in response to the claim that the VMM hypothesis was 'eliminated' by the old age of the rock data, suggested that the contrary data should be 'disregarded' so that 'everything fits just as it should' (Frankel 1982, 36).

6.2.6   The GWS Model and Early Data on the Weak Neutral Current

Our last and most extensive case study concerns the Glashow–Weinberg–Salam (GWS) model and evidence concerning its main prediction, namely the so-called weak neutral current (NC), in the early 1970s. My discussion will take its lead from a remarkable statement about the acceptance of the GWS model by one of its 'discoverers', Steven Weinberg, made in his well-known book, *Dreams of a Final Theory*, which I want to quote in full:

> One may ask why the *acceptance of the validity* of the electroweak theory was so rapid and widespread. Well, of course, the neutral currents had been predicted, and then they were found. Isn't that the way that any theory becomes established? I do not think that one can look at it so simply. . . . *what really made 1973 different was that a theory had come along that had the kind of compelling quality, the internal consistency and rigidity, that made it reasonable for physicists* to believe they would make more progress in their own scientific work by *believing the theory to be true* than by waiting for it to go away. (Weinberg 1993, 97; emphasis added)

In other words, for Weinberg, the difference-maker for the 'acceptance of the validity', i.e., the belief that the GWS model was confirmed, was *not* the discovery of the neutral currents (its main prediction) but rather the 'compelling quality' of GWS's internal properties.

Consistent with Weinberg's assessment and contrary to standard historical accounts (Galison 1983, 1987), I have argued elsewhere in detail that the evidence for the GWS's main prediction, the weak neutral current, was conflicting and ambiguous (Schindler 2014b). Physicists nevertheless cited the available positive NC evidence in support of the GWS model, neglected the negative evidence, and did not pursue alternative interpretations of the results produced by the 'NC discovery experiments'. Instead of despairing in the face of the evidential uncertainties they were confronted with, physicists looked elsewhere for reasons to believe, I suggested – namely, to the virtues of the GWS model such as unification, simplicity, beauty, external consistency, and fertility.

In order for us to develop a better grasp of these virtues, I first want to provide some background about the nature and purpose of the GWS model.

6.2.6.1   The GWS Model: A Primer

There are four fundamental forces in nature: gravitational, electromagnetic, strong, and weak forces. The recent history of physics can be viewed as an attempt to unify these forces into a single theory. One important step towards the (misleadingly coined) 'theory of everything' was taken in the 1950s–1970s with the unification of electromagnetic and weak forces into 'electroweak' forces. The development of electroweak models goes back to the early 1950s (see, e.g., Pickering 1984a; Morrison 2000), but the first lasting contribution came from Sheldon Glashow. Motivated by several analogies between photons, the mediators of the electromagnetic force, and the hypothesized mediators of the weak force, Glashow in 1961 devised an electroweak gauge theory of leptons, i.e., a quantum field theory in which the Lagrangian is partially invariant under local transformations. The gauge symmetry in Glashow's model was $SU(2) \times U(1)$, with four intermediate vector bosons (IVBs), as the mediators of force are also referred to – namely, a triplet consisting of W^+, W^-, W° and a singlet (B°). The neutral mediator of the weak interactions (later referred to as Z°) was produced by 'mixing' the neutral member of the triplet and the neutral singlet (Glashow 1961). A major drawback of this model was the fact that the masses for the IVBs were inserted 'by hand' into the Lagrangian, rendering the model not only non-rigorous but effectively also non-renormalizable.[17] Progress was made six years later when Steven Weinberg and Abdus Salam, independently of each other (but Weinberg slightly earlier than Salam), developed a model in which the masses for the IVBs would be produced through *spontaneous symmetry breaking*[18] from the four massless IVBs assumed in Glashow's model. In Weinberg's model, these IVBs gained their masses through the

---

[17] Renormalizability is the property of a theory by which certain types of infinites occurring in higher-order approximations in those theories are eliminable by replacing them with measured values (e.g., electron mass and charge).

[18] In spontaneous symmetry breaking (SSB) the field theory Lagrangian possesses a symmetry that the described physical system, *on the face of it*, does not possess. A well-worn standard example of SSB is ferromagnetism. A ferromagnet such as a bar magnet is made up of spinning particles which all align in one direction. However, the Lagrangian for interacting spin particles is 'rotationally invariant', i.e., it is symmetric and exhibits no preference for any one direction. It is therefore assumed that the observed physical state (displaying no symmetry) comes about from a symmetric state after SSB at the critical temperature.

Higgs mechanism, which postulates a set of scalar bosons at high energy ranges, namely an isospin doublet H$^+$, H$^\circ$, and its antiparticles H$-$, $\overline{\text{H}}^\circ$, which themselves gain mass through mutual self-interaction. When spontaneous symmetry breaking occurs, the masses of the H$^\pm$ and the $\text{H}^\circ - \overline{\text{H}}^\circ$ pair are absorbed by the W$^\pm$ and the Z$^\circ$, respectively. Since the photon absorbs no mass, a massive Higgs boson remains, which has only very recently been observed (March 2013). The only parameter left free in Weinberg's model (besides the mass of the Higgs boson) is the so-called Weinberg angle (θ_w), which determines the 'mixing' of the initial neutral vector bosons W$^\circ$ and B$^\circ$ to yield a massless photon and a massive Z$^\circ$ boson.

An immediate challenge the GWS model faced after its inception was the question of whether it would be renormalizable. In his proposal of the model, Weinberg had speculated that it would be but was not able to prove it. This was one of the reasons why the model was initially more or less ignored by the physics community (Koester et al. 1982). It took several years until the young physicist 't Hooft managed to come up with a rather complicated version of the required proof ('t Hooft 1971) that was subsequently simplified (cf. Pickering 1984a).

The clearest empirical prediction of the GWS model, which distinguished it from the then-prevalent V-A theory of weak interactions by Feynman and Gell-Mann (1958),[19] was the existence of weak neutral currents as mediated by the Z$^\circ$ IVB (see Figure 6.5). Since there were no detectors with energies high enough to detect the Z$^\circ$ particle (this became feasible only in the early 1980s), physicists set out to provide evidence for the existence of the Z$^\circ$ particle *indirectly*, as it were, through the detection of weak neutral currents (NC), which, if they were to exist, would leave characteristic signatures in particles interactions. Initially, however, it looked as though neutral currents were non-existent.

A well-known form of weak interaction in the 1960s, the hadronic decay of K-mesons, showed practically no signs of NC. These kinds of weak interactions involved only particular types, namely ones in which the 'strangeness' of the involved particles would *change*.[20] It was generally

---

[19] According to the V-A theory, current–current weak interactions are a mixture of vector and axial (i.e., V-A) vector parts, leading to the observed parity violation in weak interactions. In the first theory of weak interactions, Fermi had assumed that they had only vector character.

[20] Strangeness is a quantum number that was introduced as a 'bookkeeping device' to accommodate the 'strange' fact that kaon and lambda particles were produced at very high rates in particle collisions but decayed very slowly.

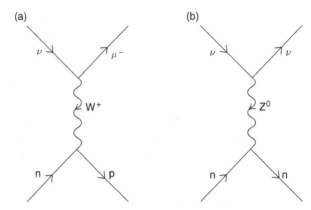

Figure 6.5 Neutrino-neutron interactions: (a) charged current event, mediated by a W+ boson, which carries a positive charge from the reaction v → μ– to the reaction n → p (where v = neutrino, n = neutron, p = proton, μ– = muon); (b) neutral current event, mediated by an electrically neutral Zo boson, where incoming neutrinos (upper left edge) retain their identity after interacting with a nucleon.
Adapted from Pickering (1984a). Reprinted by kind permission of Chicago University Press.

assumed that strangeness-conserving interactions would behave no different than strangeness-changing interactions, and there appeared to be little interest in establishing this belief experimentally (Galison 1987). This of course changed dramatically with the NC prediction by the GWS model. Although the proponents of the GWS model could help themselves to the so-called GIM-mechanism when extending the model from leptons to hadrons, invented by Glashow Iliopoulos and Maiani in 1970 to suppress strangeness changing currents,[21] it still required the existence of strangeness-conserving NC.

Finally in 1973–1974, experiments at CERN and the National Accelerator Laboratory (now Fermilab) reported evidence in favour of the NC, but the evidence was far from being unequivocal (Galison 1987; Schindler 2014b). The NAL experiments first reported a negative result and then produced a positive result under dubious circumstances. Subsequently, after improving their apparatus, they even reported

[21] The GIM mechanism was invoked to save the then-prevalent theory of weak interactions – the V-A theory by Feynmann and Gell-Mann (cf. fn. 19) – from being refuted by the absence of strangeness changing NC in kaon decay. Although the V-A theory did not postulate a neutral IVB, the exchange of a W⁺ and a W⁻ could mimic a Z° in higher orders of perturbation theory (Pickering 1984b, 183).

a result that was in fact inconsistent with the GWS model. And, most curiously perhaps, both CERN, NAL, and a third collaboration from Caltech reported results that converged on a value for the GWS model's Weinberg angle, which was significantly too low (I have called this a 'fake convergence'; Schindler 2014b). I will not go into any further detail of the experiments here. Instead, I will focus on the non-empirical virtues of the GWS model, which I believe played a crucial role of in the acceptance of the GWS model *despite* the dubiousness of its empirical support.

6.2.6.2   *The Virtues of the GWS Model*

Unification.   According to Morrison (2000), who provides a comprehensive discussion of the electroweak unification in her book *Unifying Scientific Theories*, '[u]nity was not the goal or even the motivating factor' in the development of electroweak gauge theories by Glashow and Weinberg, '[u]nity was simply not mentioned as a factor in the theory's acceptance, either by its founders or by their colleagues' (125), and '[t]he fact that a particular model had succeeded, to some extent, in producing a unification was not compelling in any sense of the word' (134). I do not agree with any part of this assessment. Morrison bases her claim that unification was not a motivating factor for the originators of the GWS model on a personal email correspondence with Glashow and Weinberg. However, at least from Weinberg's ground-breaking article (Weinberg 1967), which he considered yet another attempt in the 'very long history of *attempts to unify* weak and electromagnetic interactions' dating back to the 1930s (1266), it is glaringly obvious that unification was indeed a central driving force:

> Leptons interact only with photons, and with the intermediate bosons that presumably mediate weak interactions. What could be more natural than to *unite* these spin-one bosons into a multiplet of gauge fields? *Standing in the way of this synthesis* are the obvious differences in the masses of the photon and intermediate meson, and in their couplings. *We might hope to understand these differences* by imagining that the symmetries relating the weak and electromagnetic interactions are exact symmetries of the Lagrangian but are broken by the vacuum (1264; emphasis added).

Glashow opened his contribution to the GWS model six years earlier just as unequivocally:

> At first sight there may be little or no similarity between electromagnetic effects and the phenomena associated with weak interactions. *Yet certain remarkable parallels emerge* with the supposition that the weak interactions

are mediated by unstable bosons. ... *The purpose of this note* is to seek such symmetries among the interactions of leptons *in order to make less fanciful the unification of electromagnetism and weak interactions.* (Glashow 1961, 579–80; emphasis added)

Furthermore, the *vast majority of proposed gauge theories* that were put forward in the 1970s sought to unify the two kinds of interactions. In fact, there was just a single (!) relevant theoretical approach at the time that would *not* seek to unify weak and electromagnetic interactions. And as we shall see in a moment, this approach had very few adherents. These are good grounds for thinking that unification was a factor in the theory's acceptance, regardless of what physicists may say now their reasons were.

Symmetry.   It is striking that *all* electroweak gauge theories presupposed that the relevant symmetries of nature were exact, despite appearances, and therefore incorporated a Higgs mechanism. But what are the grounds for positing such exact symmetries? Calling upon Plato's parable of the cave, Weinberg writes that although 'nature does not *appear* very simple or unified ... by looking long and hard at the shadows on the cave wall, we can at least make out the shapes of symmetries, which though broken, are *exact principles governing all phenomena, expressions of the beauty* of the world outside' (Weinberg 1979, 556; emphasis added). In other words, symmetries were posited *despite* the appearances, not because of them.

The attraction of unification and exact symmetries was strong. It was so strong, in fact, that other empirically adequate approaches that did without electroweak unification and exact symmetries were completely ignored by the vast majority of the physics community. According to Sakurai (1978, 66), there were 'only three people in the world engaged in the heresy of contemplating "alternatives" [to electroweak models] – Bjorken at SLAC, my collaborator, Pham Quang Hung, and myself'.[22] The paucity of response to these alternative approaches led to some frustration among its proponents. Bjorken (1979), for instance, after setting out his phenomenological approach which preserved 'all the predictions of the standard model for neutrino-induced neutral currents ... *without* assuming weak-electromagnetic unification, existence of intermediate bosons, or existence of a spontaneously broken local gauge symmetry' (335; original emphasis), concluded that 'it is unlikely that either the arguments or the alternatives we have sketched have enough force to induce many theorists to abandon gauge theories. There are strong, albeit *mostly subjective, reasons* favouring

---

[22] For a systematic overview of those alternatives, see Sakurai (1974).

the gauge-theory approach' (345; emphasis added). In another publication, Bjorken clarified that by 'subjective reasons' he meant 'those which persuade even in the absence of data' (Bjorken 1977, 702). Bjorken listed six such reasons: electromagnetic unification; the 'intermediate-vector-boson hypothesis' that was used to effect unification in most model;[23] 'the origin of gauge boson- and fermion mass', which in the gauge models was accomplished by SSB, i.e., 'an underlying local gauge principle'; renormalizability; and universality (Bjorken 1979, 345–6). At one point Bjorken found a rather drastic way to air his frustration about the lack of attention to his alternative approach:

> [I]t is hard to find *any* theorist (including myself) working actively and continuously on theories which lead in a direction contradictory to that of gauge theories ... To be sure, the absence of criticism of gauge-theory ideology these days is quite understandable. To work on something else is to become a bit of a social outcast, and that is something the younger (untenured) generation may choose not to face. (Bjorken 1977, 701)

The preference for gauge theories and, in particular, the preference for the GWS model went deep. So deep, in fact, that it even affected data analysis.

Sakurai (1974) had suggested that muonless events be interpreted in terms of baryonic current rather than in terms of the NC mediated by the Z° particle, as envisaged by the GWS model. In his paper, Sakurai complained, with candour rather untypical for publications in the *Physical Review*, that experimentalists 'by default' interpreted their data (i.e., muonless events) in terms of the GWS model and urged them in future to approach their data '*unhampered by theoretical prejudices*' (Sakurai 1974, 9; original emphasis). In the practices 'followed by most experimental rapporteurs in major international conferences', however, physicists assumed that 'half of the theory is correct' before comparing for consistency of the Weinberg mixing angle between NC/CC ratios for neutrinos and antineutrinos (Sakurai 1978). It was therefore 'difficult to see whether models with qualitatively different sets of coupling parameters [than the ones of the GWS model] also fit the same data' (51–2). But what were the causes of this preference for the GWS model over its electroweak gauge theory alternatives?

Simplicity. There are three aspects in which the GWS model was simple. First, the fact that the GWS model was renormalizable, in

---

[23] The only exception was a model by Georgi and Glashow (see discussion that follows).

Weinberg's words, meant that it satisfied a constraint which allowed 'only a few *simple* types of interaction' (Weinberg 1979, 516). Second, it comes equipped with 'the simplest possible' Higgs mechanism for the class of SU (2) × U(1) gauge models (Weinberg 1974, 257). Subsequent models all proposed more complex forms of the Higgs mechanism. Third, the GWS model posited no more heavy leptons than were known at the time of its proposal. With regard to the two last aspects, the GWS model had a clear edge over its competitors. Let us consider these aspects in turn.

In his Nobel Prize lecture, Weinberg explained that renormalizability was a very important constraint on his theorizing about electroweak interactions. Taking inspiration from Dirac's use of simplicity in devising quantum electrodynamics, but being sceptical about 'purely formal ideas of simplicity' in the context of 'theories of phenomena which have not been so well studied experimentally', Weinberg wrote:

> I thought that renormalizability might be the key criterion, which also in a more general context would *impose a precise kind of simplicity on our theories* and help us to pick out the one true physical theory out of the infinite variety of conceivable quantum field theories. (Weinberg 1979, 547)

Renormalizability imposes simplicity on theories because,

> [r]oughly speaking, in renormalizable theories no coupling constants can have the dimensions of negative powers of mass. But every time we add a field or a space-time derivative to an interaction, we reduce the dimensionality of the associated coupling constant. *So only a few simple types of interaction can be renormalizable.* (Weinberg 1979, 546–7)[24]

In particular, the renormalizability constraint ruled out SU(2) × SU(2) symmetry, which Weinberg had first considered for describing electroweak interactions in leptonic currents. Although Weinberg was not able to prove the renormalizability of the GWS model when he proposed it, he clearly hoped that it would be (Weinberg 1967). And as mentioned earlier, 't Hooft later did manage to prove the renormalizability of the GWS model ('t Hooft 1971). Renormalizability, however, was no unique feature of the GWS model. On the contrary, any forthcoming electroweak alternative model was going to have to satisfy this constraint. The second aspect in which the GWS model was simple was indeed unique.

When Weinberg devised the mechanism for spontaneous symmetry breaking in the GWS model, he assumed that symmetry is broken in the

---

[24] As an example of a non-renormalizable theory, Weinberg mentions the Fermi theory of weak interactions, in which 'the coupling constant has the dimensions of [mass]$^{-2}$'.

'simplest possible' way for this kind of model with a scalar doublet (Weinberg 1974, 1979). Even several years after the GWS model was put forward and after a number of other electroweak gauge models had been developed (mostly $SU(2) \times U(1)$ alternatives; cf. Bernabeu and Jarlskog 1977), it was, according to a comprehensive overview article, 'among all the gauge models proposed so far . . . [still] *by far the simplest model incorporating unification*', for it contained the 'simplest Higgs mechanism' (Sakurai 1978, 64; emphasis added).

Before the neutral current search in neutrino scattering experiments were announced at CERN and Fermilab, probably the most attractive competitor to the GWS model in the early 1970s, as indicated by citation analyses (Koester et al. 1982), was the model proposed by Georgi and Glashow (1972) (not to be confused with the Grand Unified Theory, GUT, by the same authors, Georgi and Glashow 1974). And interestingly, it is the *only* competitor model mentioned by *both* Glashow and Weinberg in their respective Nobel Prize lectures. There was indeed a sense in which the Georgi–Glashow (GG) model, which was based on a $SU(2)$-gauge symmetry group only, was more attractive than the GWS model. It provided a 'more profound unification' than the GWS model, for it classed the neutral photon and the two charged massive vector bosons (W^+, W^-) in the *same* family of particles (cf. Weinberg 1974; 't Hooft 1980) and it accomplished the electroweak unification without the postulation of a new neutral IVB (i.e., the Z° in the GWS model). But the GG model had several weaknesses regarding its coherence (as defined in Chapter 5) and simplicity. First, contrary to the GWS model, the electron mass was presumed to be the difference between two mass terms, one being bare mass and the other being generated by spontaneous symmetry breakdown. Bjorken pointed out that '[n]o rationale for the miraculous cancellation is given', rendering the model 'utterly unbelievable'. In order to render the plausibility of this model 'at least non-vanishing', one had to introduce another Higgs particle into the model (Bjorken and Llewellyn Smith 1973, 33–4). In the GWS model, in contrast no such miraculous cancellation had to be assumed and, as already mentioned, only one Higgs particle was required (Weinberg 1967). Furthermore, several authors considered the model 'artificial' (Bjorken 1972) or even 'ugly' (Glashow 1980), which probably had to do with its generalization to hadrons, which, in Georgi and Glashow's own words, contained 'considerable arbitrariness' (Georgi and Glashow 1972). Third, and perhaps most importantly, the GG model was epistemically much more costly than the GWS model in terms of the number of particles it invoked: it required the existence of altogether five quarks and not less than four new heavy leptons. In comparison, the GWS model made do with four quarks and no new leptons.

External consistency.   The GWS model proved to be consistent with the extension from three (u, d, s) to four quarks, with the addition of a quark (c) that Glashow coined 'charm', which had been argued for in 1964 by Bjorken, Glashow, and others on the basis of symmetry between those four quarks and the four known leptons. There was not going to be evidence for charm until the mid-1970s (see further on). However, it so happened that also the GIM mechanism, invoked to suppress strangeness-changing neutral current in the GWS model, *required* the existence of charm. There were thus two virtuous models (in terms of symmetry and simplicity) in different domains which pointed to the existence of charm. In contrast, all alternatives to the GWS model postulated new quarks for which there were no independent (theoretical) reasons for belief. But these additional quarks were 'unnecessary to describe the known hadrons' and indeed would have implied new quantum numbers for strong interactions. Such additional quarks were also referred to as 'fancy' (De Rújula et al. 1974, 400). Again, the principle of parsimony appears to have weighed heavily on physicists' minds.

Fertility.   After the discovery of the J/psi-particle in 1974 (Aubert et al. 1974; Augustin et al. 1974), which gave first indications of the existence of a new quark, it took another couple of years until Goldhaber et al. (1976) provided compelling evidence that the GWS model's prediction of the weakly interacting charmed quark was actually correct.[25] Leading to several further discoveries of new elementary particles – including the IVBs of the weak force in 1983, for which there had been only indirect evidence in the 1970s – the GWS model and its extension to the strong interactions around 1979, culminating in our today's 'Standard Model', turned out to be an extremely fertile research programme. Although the fertility of the model is perhaps the most impressive virtue of the GWS model, knowledge of it was of course not available to the physicists who made a choice for the GWS

---

[25] As Pickering (1984b, 267ff.) points out, although there was evidence for hidden charmed states before 1976, this 'shed no light on the weak interactions of the constituent quark', as predicted by the GWS model, since 'the decays of the ψ and ψ' to the χs, η_c, η_c' and to normal hadrons were all dominated by the strong interaction'. There were also a number of neutrino interaction experiments that sought to generate evidence for charm, but none of them were fully convincing (Pickering 1984b, 276, fn. 28).

model and against its most promising early competitor, the GG model, in the early 1970s.

6.2.6.3   The GWS Model Concluded

The GWS model was adopted by the physics community in 1973–1974 on the basis of evidence for the weak neutral current produced by experiments carried out at CERN and NAL. Negative evidence was neglected, and experiments were rerun until they generated a positive result. Consistent with a suggestion by Weinberg, I argued that the GWS model had several virtues which gave physicists reasons for belief in addition to the evidence for the GWS's prediction of the NC. Symmetry, unification, and simplicity, in particular, are explicitly mentioned in the literature as reasons for why physicists adopted the GWS model.

6.3   The Fourth Virtuous Argument for Realism: The Argument from Choice

In the foregoing historical cases, we saw that scientists adopted theories even though there was evidence against those theories. Eddington and his collaborators saw Einstein's general theory of relativity confirmed by the light-bending data even though one out of three data sets contradicted Einstein's theory. Crick and Watson ignored Franklin's antihelical x-ray diffraction picture. High-energy physicists adopted the predecessor to the standard model, the GWS model, despite the data for its main prediction of the neural current not being compelling. And so on. On the Negative View, according to which theoretical virtues are not at all epistemic, these observations are puzzling. If empirical adequacy really is the only epistemic virtue, then – by virtue of the dictatorship condition – it seems highly irrational for scientists to adopt a theory that is not empirically adequate (at the time of its consideration). We know that these theories were adopted – i.e., believed rather than just pursued – because evidence contradicting the theories in question was questioned and dismissed, even when there was little evidence for confounders undermining the negative evidence.

The confidence scientists had in the evidence that was available *in favour* of the theories in question, I suggested, was boosted by the virtuous explanations these theories offered for this evidence. For example, Einstein's theories of relativity have great unifying power (marrying electromagnetics and mechanics), Watson and Crick's model provides a simple replication mechanism, Mendeleev identified a criterion (atomic weight) that allowed the unification of many known elements in a coherent way, the VMM hypothesis gave a simple explanation of the zebra pattern of magnetic anomalies and allowed the unification of several other observations, and the GWS model unified electromagnetic and the weak interactions in the simplest way. The virtues of these theories, however, could boost the confidence of the scientists in question legitimately only if the virtues in question were epistemic – contrary to how the Negative View would have it. I stress that this does not mean that theoretical virtues would *guarantee* the truth of the relevant theory. Instead, I believe, we should conceive of the epistemicity of theoretical virtues in terms of probability-raising, so that $P(T_{virtuous}|E) > P(T_{non-virtuous}|E)$. That is, some piece of evidence for a theory T is capable of raising the probability of T more when T is virtuous than when T is not virtuous. In contrast, if virtues were merely pragmatic, as the Negative View has it, then such probability-raising could not legitimately be affected by theoretical virtues.

The foregoing case studies suggest that theoretical virtues do (at least sometimes) boost scientists' beliefs; otherwise the scientists in question would surely not have sought to undermine those data that contradicted the relevant theory. So either the scientists in question acted utterly irrationally when they allowed their confidence in their theories to be boosted by those theories' virtues, or theoretical virtues are indeed epistemic. A good naturalist philosophy, as we shall explain in more detail in Chapter 7, ought to make as much rational sense of scientific practice as possible. The second aforementioned option, namely that theoretical virtues are epistemic, *does* offer us a rational explanation of scientists' theory choices. We should therefore accept it. I call this argument for the epistemicity of theoretical virtues in favour of realism the *argument from choice*.

6.4 Objections

In response to my argument from choice, the antirealist may raise a number of objections. First, the antirealist might insist that scientists only *pursued* the previously discussed virtuous theories (maybe just in order to explore theoretical possibility space) but did not in fact believe them. Yet the historical evidence we possess points in the opposite direction. The scientists in question strongly doubted the reliability of the data which contradicted virtuous theories. Had they merely pursued these theories, there would have been no reason for them to doubt the contradicting data, as pursuit sets very low demands on empirical adequacy. They also would have suspended judgement as to whether or not the virtuous theories in question were confirmed by the positive evidence. Yet they did consider the virtuous theories to be confirmed by the positive evidence despite the apparently negative evidence. They thus indeed believed rather than just pursued their virtuous theories.

Second, on behalf of the antirealist, one might object that the empiricist position is not undermined by the historical cases cited in this chapter, because the evidence contradicting the theories in question eventually turned out to be false. Therefore, the empirical adequacy of the theory in question was never *really* threatened by the evidence in these cases. For the argument from choice, however, it matters little whether we retrospectively can assure ourselves that something must have been wrong with that contrary evidence. What does matter is only the knowledge that was available to the scientists choosing a theory *at the time*. And at the time the situation was such that the scientists in question had no evidence for the unreliability of the contrary evidence. In fact, as explained earlier, the contrary evidence was so compelling for some scientists that they resisted adopting the virtuous theories in question. For the argument of choice to have support, this suffices: scientists adopted virtuous theories even though there seemed to be (strong) evidence against them.

Third, one may object that all my historical cases show is that the scientists opting for theoretically virtuous but empirically inadequate theories were just heavily biased towards their own theories and simply acted out of self-interest when defending their theories against counter-evidence. In general, I think we should use epistemic explanations for scientists' theory choices when they are available, rather than sociological ones. This is what we do, for example, in response to the work of the proponents of the Strong Programme in the sociology of science, who have sought to exploit the thesis of underdetermination of theories by evidence to advance their own agenda (Burian 1990). What's more, the specific sociological explanation that might be given here (acting out of self-interest) may well go the other way. That is, one may well argue that the scientists in question would have acted more wisely in their self-interest had they not questioned the reliability of the evidence contradicting their theories. This is so because these scientists took a great risk of ruining their own reputations when shedding doubt upon the negative evidence without having much concrete evidence for its unreliability. It is not plausible that they would have taken that risk had they not believed that they possessed very good epistemic reasons for believing their virtuous theories.

Fourth, one might object that I have portrayed van Fraassen's position unfairly and insist that van Fraassen is not so much interested in science methodology but rather in the right interpretation of science. Since the theories in question, despite first appearances to the contrary, eventually turned out to be empirically adequate, the empiricist picture of science is not undermined. Note, however, that van Fraassen *does* give methodological advice. In what I described as the 'dictatorship condition' in Section 6.1, for example, van Fraassen speaks about the right 'preference' of virtues and the legitimate 'advocacy' of theories. More importantly, his axiological analysis of science is *bound to* constrain methodological maxims: an end implies certain means, namely those that can successfully be used to reach that end (Laudan 1987a; cf. Chapter 7). When the end of science is empirical adequacy, then it is *rational* to use those means that further this end, in particular any means of increasing the empirical content of our theories, and it is *irrational* to use those means that counteract this end. If the Negative View is correct and theoretical virtues are just pragmatic, then adopting theories on the basis of their virtues, rather than on the basis of their empirical adequacy (at any moment in time), as in our cases, is to use means that ought to be ineffective with regard to furthering the goal of science. And a commitment to a particular future performance of a theory, as we saw, for van Fraassen, does not count as epistemic.

Finally, a more general objection one might raise against my argument from choice is that we are not obliged to make rational sense of *all* scientific practices. Instead of having to accept that theoretical virtues are epistemic, the objector could inject, we may very well accept that the historical cases we have discussed are just examples of irrational serendipity. I would consider this a biting-the-bullet type of response.

Although there are of course practices which are not rational and therefore ought not to be accounted for by the philosopher of science in rational terms, other practices do have a sound rational explanation. The practices discussed in this chapter belong in the latter category. Rejecting this explanation without some good reason why we should not embrace it would strike me as irrational itself. Furthermore, we should make ourselves aware of the fact that the Negative View not only implies that scientists were utterly irrational when they adopted virtuous but empirically inadequate theories, but it also implies that these scientists plainly made the *wrong* methodological decisions. On the Negative View, they should never have adopted those theories, as theoretical virtues are *not at all* truth-conducive on this view. This is particularly unpalatable given how much we celebrate the relevant individuals as some of the greatest scientists there have ever been.

In view of all this, the antirealist might feel compelled to concede that theoretical virtues are epistemic after all. The antirealist may however still want to insist that theoretical virtues are only epistemic with regard to a theory's observables, and not necessarily with regard to a theory's unobservables. Such a position, although inconsistent with van Fraassen's idea that commitment is no epistemic notion, would, it seems, be consistent with the antirealist's overall position (in particular, belief only in observables and in the goal of science being empirical adequacy).[26] Yet, although such a view might appear to reconcile the antirealist position with the discussed cases, this concession threatens a central tenet of antirealism itself. If the antirealist admits that scientists may adopt theories on the basis of their virtues even when those theories are not empirically adequate (at the time the theories are adopted), they thereby accept that (perceived) empirical adequacy (at some point in time) is not necessary for belief. But this surely is no option for someone who holds – as the empiricist does – that the facts are the *only* grounds we have for belief.

---

[26] Van Fraassen has explicitly rejected this option in print (with others) (Ladyman et al. 1997). Instead, he has opted for a more radical view, also known as voluntarism – for reasons entirely unrelated to the ones expressed here. More specifically, van Fraassen believes that 'what it is rational to believe includes anything that one is not rationally compelled to disbelieve' (van Fraassen 1989, 172–3), whereby the bounds of rationality are set by coherence only (van Fraassen 2000).

CHAPTER 7

Philosophy of Science by Historical Means

The history of science looms large in the philosophical arguments of this book. A legitimate, and age-old, question one might ask is: how can descriptions of historical facts possibly impinge in a meaningful way on our philosophical views, which are presumably normative? Although philosophy has indeed traditionally been concerned with normative questions, philosophical *naturalism* has long been accepted as an important strand not only within the philosophy of science, but also within philosophy more generally (Quine 1969; cf. Papineau 2015).[1] There are many readings of naturalism, but at least in one version, naturalism is the view that philosophical theorizing should be informed in one way or another by the sciences. In the philosophy of science, it is particularly hard to see how one could reasonably *not* be a naturalist: philosophy of science must engage with facts about science in order for it to be about *actual* science, rather than about some ideal construct of what science ought to be. Regardless, any naturalist faces the problem of the norm–fact divide. Clearly, for example, no matter how many people get killed in wars, killing is not made more acceptable by the fact that killing occurs frequently. Or, likewise, the norm not to fabricate one's data would not be undermined by the large amounts of scientists violating this norm. So *some* story must be told by the naturalist philosopher about how (specific) facts can impinge on (specific) norms.

In the first section of this chapter, I assess two phases in Laudan's work on the relationship of the history and philosophy of science: a more naïve one and a more sophisticated one. Neither of those, I argue, provides a satisfactory solution to the problem of the norm–fact divide. In Section 7.2, I then turn to the idea of rational reconstruction by considering the accounts of Lakatos, Kuhn, and Feyerabend. In the same

---

[1] Recently, the idea that philosophical intuitions should be subjected to 'experimental tests', by comparing them with the intuitions of laypeople, has gained much attention.

section, I introduce the idea of the *Kuhnian mode of HPS*, according to which philosophical norms about science are motivated, but not justified, by historical facts. In Section 7.3, I turn to concept clarification as another fruitful role for the history and philosophy of science.

7.1 Laudan's Naïve and Sophisticated Naturalism

Few other philosophers have reflected on the methodological relationship between the history of science and the philosophy of science as Larry Laudan. His work on this topic can roughly be divided into two phases: an early, naïve phase and a later, more sophisticated phase.

7.1.1 Laudan's Pre-Analytic Intuitions and Test Cases

In his early phase, Laudan (1977) sought to address the norm–fact divide by identifying uncontroversial cases in the history of science 'about which most scientifically educated persons have strong (and similar) normative intuitions' and which could serve as 'decisive touchstones for appraising and evaluating different normative models of rationality' (160). As examples of such cases, Laudan mentions Newtonian physics and the rejection of Aristotelian physics 'by, say, 1800' and the rejection of homeopathy and the acceptance of pharmacological medicine 'by, say, 1900' (160). Our 'pre-analytic intuitions' about such cases, Laudan urged, we must accept 'as a matter of faith' and take 'their rationality for granted' (161).[2]

It is fairly easy to see that Laudan's pre-analytic intuitions are unsuited to cut any ice in the methodological debates philosophers of science are generally interested in: they are too unspecific. For example, they will not help us arbitrate between those who believe that novel success is more valuable than accommodative success (Chapter 3) or between those who believe that theoretical virtues are epistemic and those who don't (Chapter 1). Perhaps pre-analytic intuitions do help with broader questions such as 'what is science?' or maybe there are more fine-grained but nevertheless unequivocal pre-analytic intuitions after all. Even if that were the case, there are some further problems with this approach, some of which are ironically provided by Laudan himself after a later change of mind

---

[2] This idea closely resembles the 'meta-methodology' Lakatos applies to rival methodologies (see the discussion that follows and especially fn. 8), although he appealed to the 'basic value judgements of the scientific elite' (Lakatos 1970b, 110–3) rather than to *our* intuitions about historical cases (as Laudan does). Laudan (1986) admits being inspired by Lakatos.

(Laudan 1986). First and foremost, if it is really pre-analytic intuitions against which we test methodologies, then it is questionable what kind of role history is supposed to play in the process. We might, instead, just use our intuitions about good methodology than our intuitions about historical cases. Second, as Laudan himself points out with regard to novel success, it is far from clear that our intuitions agree as nicely as they do in fairly unproblematic cases such as Newtonian physics; oftentimes our intuitions seem to be divided over important methodological issues. Third, when pre-analytic intuitions really are the 'touchstones' for methodology, it is doubtful whether methodologies could really ever have any normative force against our intuitions. But then, it seems, the constructive and critical role of methodologies is rather limited.

Instead of testing methodologies against pre-analytic intuitions about a set of historical cases, Laudan later, with a number of colleagues (Laudan et al. 1986; Donovan et al. 1988), explored the idea of testing methodologies directly against historical cases. Although this project gained some notoriety within history and philosophy of science quarters, such an approach seems highly problematic. Not only does it presuppose a highly naïve picture of empirical tests (Duhem–Quine and all), but it also – contrary to his earlier approach – simply ignores the norm–fact divide. For example, it is really hard to see how the norm that a new theory ought to conserve its predecessor's empirical success could plausibly be tested by historical cases. Surely we should not conclude that this norm is false were we to discover a number of cases where this is not the case (Laudan et al.'s 'test result' with regard to that question is mixed).

7.1.2 Laudan's Normative Naturalism

In his book *Science and Values* and culminating in the article 'Progress or Rationality? The Prospects for Normative Naturalism' (Laudan 1987a), Laudan developed a more sophisticated form of philosophical naturalism. Central to his approach is the idea of instrumental or conditional rationality: *if* one's aim is A, then one *ought to* do *m*, where *m* is the *best means* for achieving *ends* A. Whether or not *m* is a good means for achieving A, for Laudan, is an entirely empirical question. That is, has *m* been an empirically reliable means for achieving A in the past? Are there other means that did even better? Laudan puts no constraints on aims except that they must be realizable, 'for absent that realizability, there will be no means to their realization and thus no prescriptive epistemology that they can sustain' (Laudan 1990, 47). Methodological rules about science we can then construe as *theories* about the best means for reaching a certain goal. Indeed, methodological theories *about* science, for Laudan, are on a par with scientific theories themselves when it comes to their epistemic justification: 'we can choose between rival methodologies in precisely the same way we choose between rival empirical theories of other sorts' (Laudan 1987a, 24). For example, Laudan believes, we can empirically test the Popperian methodological norm 'avoid ad hoc hypotheses, if your goal is falsifiable theories', by historically investigating whether avoidance of ad hoc hypotheses has in the past resulted in more testable theories (27). This is quintessential naturalism. Contrary to the likes of Quine, however, who saw no room for normativity in a fully naturalized philosophy, Laudan believes that instrumental rationality allows for normativity and description to be reconciled. One can *describe* the aims of scientists, *describe* the means they chose, and then *judge* scientists by whether or not they chose the best means for pursuing their aims.

Another feature of his normative naturalism, which Laudan highlights, is its flexibility: it allows us to accommodate methodological changes and frees us from any kind of 'whiggish' historicism. What, for example, was rational for Newton, no longer needs to be rational for us. As Laudan explains, 'it was one of the central aims of [Newton's] natural philosophy to show the hand of the Creator in the details of his creation', and he therefore saw it as his goal to '[produce] theories which were either certain or highly probable' (Laudan 1987a, 22). Newton's theory choice must, accordingly, be judged by his own lights and not by ours (22). 'Whatever rationality is', Laudan concludes, 'it is agent- and context-specific' (21). In fact, Laudan insists that it is necessary for ascribing rationality to an agent's action that the agent herself believe that she would promote her ends by her actions. For Laudan, it makes no sense for us to say that a scientist might have acted rationally in furthering some other ends or 'the' goal of science (such as truth) without those ends being her own. He also rejects the idea that 'agents who acted effectively so as to promote *their* ends may turn out to be irrational' (23; emphasis added). Since we cannot hold scientists of the past accountable by *our* aims and norms, neither can we assess science methodologies 'by seeing whether they render rational the actions of great scientists of the past' (23), as Lakatos would have recommended it. The looming relativism Laudan regards as unproblematic, because he makes sure to divorce the question of rationality from the epistemic question of progress (and truth). Progress, Laudan concedes, *does* have to be measured against *our* standards. We can do so, Laudan suggests, by comparing the theory choices a certain methodology recommends with the theories scientists actually adopted. If the methodological recommendations would turn out to accord with the theories scientists did *not* choose, then this would be a prima facie reason for Laudan to reject the relevant methodologies (29). Of course, this implausibly presupposes that the choices scientists made are always the best possible ones and always result in progress.

Although I have no objections to construing rationality in instrumentalist terms, I do think that many of the conditions Laudan imposes on it are highly implausible. In particular, I do not believe that the relevant goals must be espoused by the agents whose rationality is assessed, that these goals be realizable so agents can take steps to actually achieve them, or even that all rationality is agent-relative.

Most philosophers of science believe that truth or empirical adequacy is *the*, or at least one important, goal of science. But truth and empirical adequacy are neither agent-relative nor are they thought to be actually realizable; they are utopian goals (cf. Chapter 1). It is also not required that scientists actually embrace these goals for them to be assessed with regard to these goals. Let us consider some of these points in more detail.

Scientists may have rather diverse personal goals, such as fame and attention, and still be legitimately judged with regard to the goal of truth or empirical adequacy. Van Fraassen (1980) once illustrated this point nicely with an analogy to chess: a chess player's own primary goal to play chess might be to intellectually challenge themselves, to socialize, or simply to have fun, but when we judge the rationality of their actions as chess players, we should of course make our judgements relative to the goal of winning the game of chess. That chess players must abide by the rules of chess and aim for the same goal is obvious. It may be less obvious in the case of science. I believe, though, that the philosopher is well advised to do as Einstein once recommended with regard to understanding theoretical physicists and their methods: 'don't listen to their words, fix your attention on their deeds' (Einstein 1982). In the terminology once coined by Lakatos, 'fix your attention to the methods *implicit* in scientists' practices, rather than to their *explicit* reflections about them' (cf. Worrall 1988).

I agree with Laudan about the importance of instrumental norms when it comes to rationality. I also don't believe there is any compelling reason not to treat epistemic rationality as a form of instrumental rationality where the goal is truth or empirical adequacy.[3] Accordingly, the respective norms would be of the form 'relative to the goal of truth or empirical

---

[3] There is an interesting discussion about whether epistemic rationality reduces to instrumental rationality. See Kelly (2003), Leite (2007), and Brössel et al. (2013). I tend to side with the reductionists.

adequacy, it is rational for agents to do m', whereby m would be the best means for achieving truth or empirical adequacy. But if that is the case, and agents' rationality can be assessed in relation to these goals independently of the historical epoch or the particular agent's goals one is considering, then it seems unjustifiable to separate the question of rationality from the question of progress, as Laudan does. Even though truth and empirical adequacy are utopian goals, we can still assess whether an agent adopts means to further these goals – contrary to what Laudan believes. For example, we can deem an agent irrational when she chooses to adopt a radically new theory that conserves *none* of its predecessor's empirical content, because this would not bring the agent any closer to either the goal of truth or of empirical adequacy.

Lastly, I want to highlight the fact that the justification of norms on Laudan's account is implausibly retrospective: a scientist acts rationally only when she follows norms which have proven effective in helping her achieve her goals *in the past*. Such retrospective justification, however, is implausible, because it would not allow us to decide not to pursue hopeless means prior to actually trying them out. For example, we surely do not need to try out the rule 'always conclude the opposite of what your evidence suggests' (as once proposed by Feyerabend) in order to understand that the rule is an irrational one. Certainly, it would border on the bizarre to claim that the rationality of logical inferences (e.g., modus ponens), so central in science, can only be determined retrospectively and only after trying them out.[4]

7.2 Rational Reconstruction

Rational reconstruction in the philosophy of science goes back to the logical positivists (Carnap 1937) but has received a lot of bad press in the hands of Imre Lakatos, in particular. Lakatos is on record for banning actual history that '"misbehaved" in the light of its rational reconstruction' to the footnotes of philosophical texts (Lakatos 1970b, 107).[5] Although there is much to be objected to in Lakatos's approach to the history and

---

[4] See Worrall (1988), Laudan (1989), and Worrall (1989a) for an interesting and entertaining exchange on the matter of changing norms.

[5] Lakatos (1978, 192) himself later described this idea as an 'unsuccessful joke' of his and claimed never to have said that this was the way 'history actually ought to be written' and that he 'never wrote history in this way except for one occasion', the exception being his *Proofs and Refutations* (Lakatos 1976).

philosophy of science, his work contains an important idea which often gets overlooked, as we shall see in a moment.

7.2.1   Lakatos and Rational Reconstruction

Lakatos was convinced – rightly so, I believe – that 'history without some theoretical "bias" is impossible' (107).[6] He was, however, ready to give normativity a role in historiography that not many historians would be happy to subscribe to, when he defined the history of science as the 'history of events which are selected and interpreted in a normative way' (108). Lakatos identified several normative views of science which he believed guided historiographic research of science at his time (explicitly or just implicitly) and coined them *methodologies*: inductivism, conventionalism, falsificationism, and his own 'methodology of scientific research programmes'. Although Lakatos nowhere defined rational reconstruction, he characterized it as the reconstruction of history in terms of a methodology with the aim of providing 'a rational explanation of the growth of objective knowledge' (91).

Lakatos distinguished between internal and external history. Contrary to the standard usage of these terms, according to which internal history is concerned only with problems internal to the dynamics of science and external history with broader societal developments and social factors, Lakatos interpreted these terms slightly more idiosyncratically (see Kuhn 1970b). Internal history, for Lakatos, is rationally reconstructed history on the basis of one of the aforementioned methodologies. External history consists of historical facts which cannot be rationally reconstructed from the perspective of the methodology in question, and for whose explanation one must resort to psychology and sociology. Lakatos was not much concerned with external history; for him, it was really just a 'rest category' for all things unaccounted for by methodology. As he put it, 'external history [is] only secondary, since the most important problems of external history are defined by internal history' (Lakatos 1978, 105). Lakatos was also upfront about internal history being 'not just a selection of methodologically interpreted facts', but 'on occasions' even 'their radically improved version' (106). As an example, Lakatos mentions electron spin, which he argued to be part of 'Bohr's research programme', since it 'fits naturally in

---

[6] For a recent discussion of theory-ladenness in the history and philosophy of science (and other issues), see Kinzel (2015).

the original outline of the programme', despite the fact that Bohr himself never pondered the idea (107; cf. Chapter 4).

Somewhat curiously, Lakatos believed that science methodologies could be tested on the basis of the internal history which they gave rise to (109). But how could they? If internal histories are constructed on the basis of methodologies, and if internal history is not just a biased collection of historical facts but sometimes even a distortion of facts, then, quite obviously, the testing procedure is either circular (when one uses the internal history reconstructed on the basis of one's own methodology to support one's own methodology) or question-begging (when one criticizes another methodology on the basis of the internal history reconstructed with one's own methodology) (Kuhn 1970b; McMullin 1974).[7]

Lakatos's writings, however, contain another, more plausible idea which often gets overlooked:

> An 'impressive', 'sweeping', 'far-reaching' external explanation is usually the hallmark of a weak methodological substructure; and, in turn, *the hallmark of a relatively weak internal history* (in terms of which most actual history is either inexplicable or anomalous) *is that it leaves too much to be explained by external history*. When a better rationality theory is produced, internal history may expand and reclaim ground from external history. (Lakatos 1978, 119; emphasis added)

In other words, methodologies can be compared and assessed by evaluating how many historical facts they are capable of reconstructing as internal history; the more facts would land in the bucket of 'irrational' external history, the worse the methodology. Not surprisingly, Lakatos claimed that his own methodology of research programmes 'turns many problems which had been external problems for other historiographies [and therefore methodologies] into internal ones' (Lakatos 1978, 104). For example, Lakatos repeatedly stressed the fact that whereas Popper's falsificationism renders scientists irrational who do not refute a hypothesis when it has counterinstances, his methodology makes it rational for scientists to keep working within a research programme *despite* anomalies so long as the research programme does not become 'degenerative', i.e., so long as it does not cede to produce novel success (114–8; cf. Chapter 4). Lakatos also

---

[7] Even more curiously, perhaps, Lakatos suggested that the testing procedure to be applied for falsificationism, inductivism, and conventionalism should be *falsificationist* (with the basis for falsification being provided by the 'basic value judgement of the scientific elite' (110) but exempted his own methodology of research programmes from such tests. Lakatos neither explains why falsificationism should be the preferred 'meta-methodology', nor does he provide any reasons for exempting his own methodology (116).

believed that his account could explain why theory change is rational, whereas Kuhn's and Feyerabend's accounts couldn't (118).[8] Although Lakatos thought that methodologies could be compared on the basis of how many historical facts they could rationally explain, he at the same time was aware that no methodology would ever be able to reconstruct *all* historical facts as internal history, as 'no set of human judgements is completely rational and thus no rational reconstruction can ever coincide with actual history' (116). Thus, Lakatos thought that a good methodology of science would seek to maximize the set of facts subsumed under internal history and minimize irrational external history, even though he did appreciate that the reconstruction of all historical facts about science was unattainable.

I think the idea of the maximization of rational historical facts (without the distortion of facts) is an appealing one – again, with the understanding that it would of course be naïve to believe that *all* practices of scientists must be rationally reconstructed. In fact, the fourth virtuous argument for realism, which I developed in Chapter 6, relies on the plausibility of this idea. Objections to this presupposition I will address in Section 7.2.3.

7.2.2 Kuhn on HPS and the Kuhnian Mode of HPS

Given his role in the promotion of the history and philosophy of science, it is rather curious that Kuhn even denied at one point that the history and philosophy of science were compatible disciplines (Kuhn 1977b). Contrary to what he was often accused of, however, Kuhn was himself interested in accounting for scientific practices in a rational way; he committed himself to the endeavour of 'rational reconstruction' (Kuhn 1970c, 236). What Kuhn was keen to emphasize was that the history of science had the potential to *transform* our view of what is rational. Defending himself against some of this kind of criticism by Lakatos, Kuhn wrote:

---

[8] As Kuhn (1970b) pointed out, it is in fact questionable whether that really is the case. For Kuhn, a paradigm crisis was not just a 'psychological crisis' in the scientific community, but such crisis was underpinned by a significant increase in the number of anomalies a paradigm was facing. Kuhn did not further specify what amount of anomalies might be significant enough to cause crisis. However, something analogous is true also for Lakatos's account. That is, it is not clear at all on Lakatos's account *when* it is rational for scientists to change one research programme for another. Lakatos even allowed for scientists to hang onto research programmes that fail to produce any empirical progress for several decades (as in the case of the research programme of Copernicus; Lakatos 1978, 184). To criticism, Lakatos responded famously, 'It is perfectly rational to play a risky game: what is irrational is to deceive oneself about the risk' (Lakatos 1970b, 104, footnote). To the extent that this is a good defence of the Lakatosian account, it is one for the Kuhnian too.

Scientific behavior, taken as a whole, is the best example we have of rationality ... That is not to say that any scientist behaves rationally at all times, or even that many behave rationally very much of the time. What it does assert is that, if history or any other empirical discipline leads us to believe that the development of science depends essentially on behavior that we have previously thought to be irrational, then we should conclude not that science is irrational but that *our notion of rationality needs adjustment here and there*. (Kuhn 1970b, 144; emphasis added)

In a reply to his critics elsewhere, Kuhn expanded by saying that his 'confidence in that theory [of his] derives from its ability to *make coherent sense of many facts which, on an older view, had been either aberrant or irregular*' (Kuhn 1970c, 236; emphasis added). Although Kuhn does not himself provide illustrations of these points in these passages, examples can easily be found in *The Structure of Scientific Revolutions* in particular. Consider, for example, Kuhn's idea of normal science.

Normal science, for Kuhn, basically consists of 'puzzle solving activity': a science's paradigm provides a theoretical and practical framework, which not only guides scientists in the choice of their problems, but also nourishes in them the expectation that the theoretical and empirical problems they are concerned with *do have a solution* within the confines of the paradigm. Having a paradigm, for Kuhn, is a large part of the reason why science is so *efficient* in solving (oftentimes, as Kuhn put it, rather 'esoteric') problems. In fact, Kuhn even defended as a necessary condition for demarcating science from non-science that a scientific community possess a paradigm (Kuhn 1970a). Now, because of the inherent limitation of any paradigm (but also because of simple practical reasons), no paradigm will ever solve all of the puzzles that can be posed within it. And on occasion there will also be observations that fall outside of the confines of the paradigm which the paradigm provides no means for accommodating. On Kuhn's concept of normal science, it is rational for scientists not to reject the paradigm in these instances, because the primary goal of normal science is the *efficient* solution of puzzles. Letting counterinstances halt the puzzle-solving process in order to consider alternatives to the paradigm would *prevent* science from being efficient. It is therefore *rational* to leave aside anomalies that resist puzzle-solving efforts and focus on those problems which the paradigm does provide effective tools for. We can indeed formulate this as a norm: leave aside those puzzles that resist solution, and focus on those puzzles that are resolvable within the paradigm in order to ensure efficient puzzle-solving.

So here we have Kuhn's aforementioned ideas concerning historically informed philosophy of science in action. The Popperian norm implies that scientists are irrational when they don't give up their falsified theories The Kuhnian norm, in contrast, is consistent with the descriptive fact that theories are regularly faced with anomalies and that scientists are normally not too disturbed by this. The Kuhnian norm, therefore, is consistent with a broader range of historical facts, whilst at the same time making good rational sense of these facts. By Lakatos's lights of maximizing rational historical facts, Kuhn's account is clearly to be preferred over Popper's. And in contrast to Lakatos, Kuhn provides a rationale for why it is acceptable that scientists leave aside anomalies. Whereas Lakatos *allows* scientists to set aside anomalies so long as the research programmes they are working within is progressive (see Chapter 4), Kuhn actually *explains* why it is rational for scientists to do so: it allows more efficient puzzle-solving.

It is furthermore worth pointing out that although the Kuhnian norm is clearly motivated by historical facts, it is not justified by them. This can be seen when stating the relevant counterfactual conditional. Imagine a world in which scientists would *never* encounter *any* puzzles resisting resolution within their paradigm; all puzzles would be solvable within a reasonable amount of time. Even in such a world, the norm of leaving aside puzzles that resist resolution in order to increase the efficiency of science would be perfectly rational. Therefore, the justifiers of philosophical norms such as the norm to leave aside recalcitrant anomalies cannot come from the practices governed by these norms. I surmise that the relevant justification, whatever it might look like in detail, will be largely an a priori justification. Let me illustrate this by comparing normal science to the following scenario.

Suppose I have several tasks that I want to get finished by the end of the week. I want to tidy up my flat, do some work in the garden, get my car fixed, see my doctor, mark essays, and finish a paper. I have no preference for any of those tasks; they all need to be carried out as soon as possible. However, my paper-writing turns out to take longer than anticipated. I have only two options: I either keep working on this task but then have less time for the other tasks and run the risk of not finishing those tasks either. Or I leave the paper aside and have a good chance of getting all the other tasks done by the end of the week, just as I had planned. Given that I have no preference for any one of those tasks and given that they all need to be done as soon as possible, the second option clearly is the more rational one for me. And this is really just analogous to the normal science case.

In both cases it is considered what one ought to do when one is trying to tackle a bunch of tasks with a *limited amount of resources* whereby (i) one task is taking longer than anticipated and (ii) the rest of one's tasks are very much manageable. The fact that it is rational to focus on the manageable tasks cannot be undermined by anything we might learn about the world. For example, I might fall sick in the middle of the week and won't finish any of my tasks. As a matter of fact, my following the norm of leaving aside unmanageable tasks would not have helped me in this case to reach my goal. But clearly this should not lead me to question the rationality of the norm. It seems most plausible that the right justification for the structure that underlies both normal science and this toy example is grounded a priori in considerations regarding utility maximization.[9]

Based on these considerations, I propose that in the history and philosophy of science, norms about science are discovered through historical analysis, but not justified by historical facts. I have elsewhere called this view the *Kuhnian mode of HPS* (Schindler 2013a) – although I do not want to claim that Kuhn himself would have thought about HPS along these lines (he very likely didn't):

- *Discovery of norms:* Historical facts about scientists' behaviour motivate the construction of scientific norms under which those actions come out as rational.
- *Justification of norms:* Historically motivated scientific norms provide constraints for rational behaviour even in worlds in which the facts which motivated these norms (in the actual world) are different. Historical facts are therefore not the justifiers of historically motivated norms.

Accordingly, we can form two counterfactual conditionals:

- *D-counterfactual:* Had there not been historical fact *h*, there would have been little reason to propose norm *n* under which *h* comes out as consistent with rational behaviour.
- *J-counterfactual:* Even if *h* had not happened, norm *n* would still provide constraints for rational behaviour.

---

[9] Herbert Simon's satisficing model is often considered a semi-empirical model of rational choice (Giere 1985). Although rational choice is relativized to the agent's *actual* interests and limitations, what is rational or not has nothing to do with the agent's actual actions. The agent may for example consistently act against her interests and goals. Such behaviour would be irrational also on that model. In other words, even though the *constraints* on rational action may well be empirical, whether or not an action is rational is not to be determined empirically.

It is easy to see that the Kuhnian mode of HPS allows us to respect the norm–fact divide, whilst at the same time taking very seriously historical practices and letting those inform our views about what we consider rational scientific behaviour. Again, although Kuhn probably would not have identified himself with the Kuhnian mode of HPS – he for example rejected the distinction between the contexts of discovery and justification – I do think it provides a good reconstruction of how Kuhn reached at least some of his insights into science.

The Kuhnian mode of HPS is not specific to Kuhn's own work; it can also be found beyond it. Consider novel success. As we saw in Chapter 3, Worrall has provided ample historical evidence that scientists appear not to value temporally novel success over accommodative success. Worrall did not conclude that scientists are irrational because our intuitions tell us that temporal predictivism is true. Instead, Worrall sought to revise predictivism in such a way that it would make it rational for scientists not to prefer temporally novel success to accommodative success (of the use-novel kind). According to Worrall's non-empirical reasoning, evidence ought to be use-novel, because evidence that is used in the construction of a theory is equivalent to the theory being ad hoc, which it ought not be.

In fact, the divide et impera moves discussed in Chapter 2 might be considered yet other examples of the Kuhnian mode of HPS. As we mentioned previously, one of the most popular positions employing a divide et impera move is structural realism. Structural realists have argued that we have good reasons to believe in the structure, but not the 'content', of scientific theories. Put slightly differently, given the history of science and radical theory change, structural realism implies that it is *rational* for us to believe in the structure of theories and not in their content. Conversely, it is irrational for us to believe in more than the structure of our theories. Of course, the rationality of the divide et impera move is different from the rationality of normal science and novel success: it is about *our* (i.e., philosophers') rationality, not the ones of scientists; and it concerns our *belief*, not our behaviour. Still, just like in the invention of the concept of normal science, history has given us an incentive to revise our prior views of rationality. More specifically, history has prompted us to revise the standard realist view and the norms associated with it (e.g., believe that an empirically successful theory is true). And once more, the justification of those revised norms does of course not come from the history that motivated that revision (but rather from the application of IBE and NMA).

Let us finally point out that the Kuhnian mode of HPS, as presented here, contrasts quite sharply with Laudan's normative naturalism with regard to the way norms are justified. On Laudan's view, all norms are instrumentalist and are to be justified empirically: means must be empirically effective in achieving ends for them to have normative force for agents. According to Laudan,

> Whether our methods, conceived as means, promote our cognitive aims, conceived as ends, *is largely a contingent and empirical question.* What strategies of inquiry will be successful *depends entirely on what the world is like,* and what we as prospective knowers are like. One cannot settle a priori whether certain methods of investigation will be successful instruments for exploring this world, since whether a certain method will be successful *depends on what the world is like.* (Laudan 1987b, 231; emphasis added)

On the Kuhnian mode of HPS, by contrast, the justification of norms about science is decidedly not dependent on the *actual* state of affairs. That is, for something to be a norm on the Kuhnian mode of HPS, it is not required, contrary to Laudan's account, that certain means have in fact proven to be effective for reaching certain goals. Instead, the rationality of norms on the Kuhnian mode of HPS presumably is to be justified a priori.

Let me conclude by emphasizing once more that in the examples we have discussed (normal science, use-novelty, and structural realism), we appear to be able to accommodate a larger number of historical facts than before by revising our views of what is rational – just as Kuhn and Lakatos would have had it. This is not by giving up on rationality, but rather by devising accounts under which the practice in question comes out as rational after all. This endeavour is at the heart of the Kuhnian mode of HPS.

7.2.3 *Anything Goes?*

The project of rational reconstruction along the lines suggested by Lakatos and Kuhn, in particular, presupposes that rationality is a virtue and that a good account of science is one that maximizes rationality. Paul Feyerabend's famous slogan of 'anything goes' seems to undermine both of these assumptions. In his *Against Method*, Feyerabend argued that 'given any [methodological] rule, however "fundamental" or "rational", there are always circumstances when it is advisable not only to ignore the rule, but to adopt its opposite' (Feyerabend 1975, 7). For example, Feyerabend explained that contrary to some prominent and intuitive methodological norms,

... there are circumstances when it is advisable to introduce, elaborate, and defend ad hoc hypotheses, or hypotheses which contradict well-established and generally accepted experimental results, or hypotheses whose content is smaller than the content of the existing and empirically adequate alternative, or self-inconsistent hypotheses, and so on. (7)

Amongst the historical cases Feyerabend adduced to back up this claim are the Copernican system's apparent conflict with some observations, the apparent inconsistency between Newtonian mechanics with Galileo's law of free fall (the latter implies that acceleration is constant, whereas the former that it decreases with the distance from the centre of the earth), Newtonian mechanics with Kepler's laws (the former implies deviations from perfect elliptical orbits), Bohr's apparently inconsistent model of the atom, and the 'refutation' of the apparent truth of the phenomenological second law of thermodynamics (entropy always increases) by the discovery of the Brownian motion of molecules (which are in *apparent* perpetual motion). Feyerabend argues that scientific progress would have been seriously stifled if scientists had abided by the letter of methodological imperatives of 'do not devise inconsistent theories', 'do not construct ad hoc hypotheses', 'form inductions (rather than counterinductions) on the basis of your evidence', etc. Thus, in view of the scientific achievements the violation of such norms has led to, Feyerabend argued that the plurality of methods is 'absolutely necessary for the growth of knowledge' (7). Successful participation in the scientific activity, Feyerabend therefore suggested, requires the scientist to be 'a ruthless opportunist ... who adopts whatever procedure seems to fit the occasion' (2).

Despite what this may suggest, Feyerabend did not believe that there were *no* methodological rules in science (231). Instead, he simply expressed aversion against what he called 'law and order science' (1), i.e., too strict or universal methodological rules (4); rigid traditions; and a fixed 'theory' of rationality (11) – contenting himself with having argued for the conclusion that 'all rules have their limits' (231). Nevertheless, his *methodological anarchism*, as he liked to call his own position, was quite far-reaching in that it invited 'experts and laymen, professionals and dilettanti, truth-freaks and liars ... to participate in the contest and to make their con-tribution to the enrichment of our [scientific] culture' (14). He even believed that voodoo magic 'can be used to enrich, and perhaps even to revise, our knowledge of physiology' (28 f.). Thus, it would probably be fair to say that Feyerabend was a sceptic not so much about rationality per se, but rather about strict and rigid rationality. His concept of rationality and his view of acceptable scientific methods were very liberal indeed.

Even though Feyerabend was very happy to advertise the breaking of norms, he never specified what made it 'advisable' to break norms (7), under which conditions a counter-rule becomes 'beneficial for science' (17), or why the breaking of a norm might mean 'progress' for science (7). There is, however, an implicit criterion Feyerabend uses. Why was it *good* for Newton to propose his theory even though it was strictly inconsistent with Galileo's law of fall and Kepler's laws? Why was it *good* for Copernicus to defend the sun-centred system even though he could not present a plausible terrestrial physics and even though the stellar parallax shift entailed by it, due to poor telescopes, could not be observed during his time? The answer is because both Newton's theory and Copernicus's turned out to be empirically successful. Feyerabend, in his rejection of methodological norms, thus implicitly subscribes to the idea that the goal of science (minimally) is empirical success. Only then does it make sense to approve the violation of norms that have as their goal empirical success. If one admits that much, however, one also commits oneself to a certain view of rationality: a scientist is rational when she takes means (such as the violation of other norms) to achieve the goal of empirical success. Again, Feyerabend thus seems to be much less radical than he might first appear (for other aspects see Brown and Kidd 2016).

A much more radical interpretation of Feyerabend's slogan 'anything goes' than Feyerabend himself seemed to have endorsed would be to deny that there are *any* scientific methods and that there is *any* rationality in science. Such a view, however, should strike the reader as outright absurd. If there were no methodological norms, there would be no (rationally compelling) reason, for example, not to cook up one's data and to plagiarize, i.e., activities which obviously undermine scientific progress.

There is yet another, more moderate view of methodological norms which nevertheless takes Feyerabend's scepticism seriously. On this moderate view, methodological norms are not categorical, but qualified, or ceteris paribus, norms. For example, the norm not to devise ad hoc hypotheses should be understood not in the categorical sense of 'you must *never* use ad hoc hypotheses', but rather in the sense of 'you ought not to use ad hoc hypotheses unless X', where X are conditions such as 'empirical success would be compromised'. Thus, there might be other concerns against which the concern articulated in the current norm must be weighed. This is the type of norm we recommended in Chapter 5 in our discussion of ad hocness.

On this moderate view, methodological norms have of course only qualified prescriptive power. If it is not categorically bad to use ad hoc hypotheses, their use should accordingly not always be chastised. In fact, I think it is questionable whether philosophy of science should have any business in prescription in the first place. A philosopher's methodology of science, I believe, should not be understood as a prescription but rather as an *expectation* of scientists to behave in certain ways in the way that scientific theories tell us that we ought to observe P, given certain boundary conditions. If we should realize that scientists are not doing X, we would not necessarily prescribe them to do X, but we might be compelled (but need not be) to change our methodology of science so that X comes out as rational – just as the Kuhnian mode of HPS would suggest.

7.2.4   *Rational Reconstruction Concluded*

Although rational reconstruction is regularly associated with the distortion of history, it can play a much more positive role in the philosophy of science. In fact, Lakatos, who is largely responsible for the bad reputation of rational reconstruction, suggested a much more plausible goal for it: philosophy should not distort the historical facts, but rather maximize the number of historical facts that can be explained in a rational way by a philosophical methodology. This idea, as we have already seen, seems to also be embraced by Kuhn, and it is part of what we referred to as the *Kuhnian mode of HPS*. The Kuhnian mode of HPS avoids the norm–fact divide, we argued, because norms about science are motivated by the historical facts, but not justified by them. The norms devised in this way need not be categorical ones; they may also be qualified norms. That is, certain norms (such as to avoid ad hoc hypotheses) might be weakened or even suspended in case other more important norms (such as empirical success) would be compromised in the pursuit of those norms.

7.3   Concept Clarification

Rational reconstruction along the lines just set out is one way to avoid the norm–fact divide when one seeks to employ historical methods for philosophical ends. Another important way is concept clarification. This will be our topic for the remainder of this chapter.[10]

---

[10]　I do not wish to suggest that these two roles are the only fruitful ones for the history and philosophy of science (HPS). Chang (2004), for example, proposes that HPS may be viewed as *complimentary science*, which uncovers and highlights long-forgotten research questions and problems that have fallen by the wayside in the pursuit of (Kuhnian) 'normal science', with the goal of exploring some of those questions with current technological means and understanding.

Philosophical concept clarification looms large in many philosophical endeavours. It was famously explored by Rudolf Carnap under the label 'explication', particularly in his *Logical Foundations of Probability* (1950). Carnap distinguished between explicandum, namely the concept that stands in need of clarification, and explicatum, i.e., the concept that is supposed to illuminate the explicandum by exhibiting a higher degree of precision and clarity. Carnap identified four desiderata for successful explication: similarity, exactness, fruitfulness, and simplicity. With regard to *similarity*, Carnap advised, a good explicatum should be similar to the explicandum so that 'in most cases in which the explicandum has so far been used, the explicatum can be used' (7). An *exact* explication, for Carnap, required the provision of 'rules of its use (for instance, in the form of a definition)'. Explication would be *fruitful*, for Carnap, if it resulted in a concept that would be 'useful for the formulation of many universal statements' in its field of inquiry (Carnap was mostly concerned with the empirical sciences and mathematics). *Simplicity*, Carnap said, was to be 'measured, in the first place, by the simplicity of the form of its definition', although he considered simplicity of 'only secondary importance' and relevant only when one had to choose between several equally fruitful concepts (7). These desiderata are best illustrated with one of Carnap's examples.

The concept of temperature is an explication of our intuitive concepts such as 'warm' or 'cold'. The two explicanda are vague and subjective in that they have no precise conditions for application and vary not only from person to person but also for the same person (room temperature can appear warmer after one has spent some time out in the cold). Clearly, the desideratum of exactness can be met by the explication of the explicanda: we can use thermometers to determine temperature. The concept of temperature is also fruitful, as it is indispensable in many scientific experiments, and arguably also simple (e.g., it's one-dimensional). The explicatum is also similar to the explicandum, but only to a certain degree: in many (but not all) cases, 'if x is warmer than y . . ., then the temperature of x is higher than that of y'.

Schupbach (2017) points out that Carnap set greater store particularly in the fruitfulness of the explicatum than he did in the similarity between the explicatum and the explicandum. With regard to the previous example, Carnap wrote that 'it would be possible but highly inadvisable to define a concept Temperature in such a way that x and y are said to have the same temperature whenever our sensations do not show a difference' (Carnap 1950, 13). Although similarity clearly is a desideratum, Carnap thought it should be discarded when the other desiderata are at stake in the explication for the sake of building more exact and fruitful concepts in the pursuit of knowledge. Carnap's approach has therefore also been described as 'concept engineering' (cf. Schupbach 2017). This approach to concept clarification is arguably to be found in many current discussions in the philosophy of biology, involving concepts such as species, organism, population, and fitness, where the primary purpose seems to be to engineer concepts 'that would yield a more fruitful biological theory' (Schupbach 2017).

Carnap's concept engineering can be contrasted with Kemeny and Oppenheim's (1952) approach to concept explication (see Schupbach 2017). Rather than engineering concepts with the primary goal of improving the fruitfulness of the concept, Kemeny and Oppenheim set greater store in similarity. The goal of their approach, as Schupbach puts it, was to 'illuminate rather than to replace the [explicandum] concept', in order to ensure clarifying a concept 'which scientists commonly apply' (Schupbach 2017, 678). Schupbach associates concept clarification, a la Kemeny and Oppenheim, with work in 'formal philosophy' on concepts such as probability, causation, causal strength, confirmation, coherence, and explanatory power. Here, Schupbach claims, the overriding concern is that the explicatum reflect 'natural or obvious' intuitions about the explicandum (681). He also takes the advancing of counterexamples to formally explicated concepts as evidence for these endeavours adhering to the Kemeny–Oppenheim approach, since 'the point of these counterexamples is to show that the proposed explicatum does not have various properties that we naturally associate with the explicandum' (682).

Although I am sure that similarity sometimes plays a bigger and at other times a smaller role in concept clarification, I am less sure that the philosophy of biology and 'formal' approaches within the philosophy of science can be as clearly associated with the Carnapian and the Kemeny–Oppenheim approaches, respectively. Philosophers of biology very much care about the descriptive adequacy of the concepts they develop. For example, discussions about the gene are *all* about finding a definition which captures all of the various uses of the gene concept by geneticists. There seems very little appetite for engineering gene concepts. Counterexamples also loom large. On the other hand, explications of the concept of causation in what Schupbach calls 'formal philosophy', for example, seem to prioritize a desideratum more valued by Kemeny and Oppenheim – namely fruitfulness of the explicanta they develop, in order to solve various puzzles that have been amassed over the years (e.g., preemption cases). If anything, a major difference between the philosophy of biology and formalistic explications is that the latter normally rely on philosophers' intuitions about the concept in question, rather than on what scientists might mean by them. Neither Carnap nor Kemeny and Oppenheim seemed to have been much interested in in the latter. In contrast, naturalistic philosophers are generally rather unsympathetic to intuition-guided approaches to concept clarification. The historian–philosopher Hanson (1962) once expressed this sentiment in a way that I'm sure many naturalist philosophers would be happy to agree with:

> To the historian such philosophy is often unilluminating because it does not enlighten one about any thing: nothing in the scientific record book is treated in such symbolic studies. (Hanson 1962, 582)

Although it would probably be unfair to say that contemporary formal approaches are as exsanguinous with regards to scientific practice as Carnap's writings were, it is still true that concerns about actual scientific practice have a rather low priority. On the other hand, Hanson probably also hits a nerve in his criticism of those who place descriptive accuracy above concept clarity:

> To the philosopher, histories of science are often unilluminating because, as a result of their chaotic diffuseness, they never reflect monochromatically: only spectra of concepts and arguments result. (1962)

I believe there is a middle ground between these two extremes of intuition-guided concept clarification and historical adequacy. That middle ground consists of explicating the concepts used by scientists. There are precedents, such as the explication of the concept of scientific models (Frigg and Hartmann 2012), and, perhaps to a slightly less practice-oriented extent, explanation, law, and theory.[11]

Seeking to ensure descriptive adequacy of the clarified concepts does not necessarily square only with the Kemeny-Oppenheim approach. As Achinstein (1974) pointed out, even the Carnapian approach requires that we figure out what is 'wrong' with the explicandum (e.g., vagueness) before replacing it with a more exact explicatum: 'but to know what is wrong with the present concept involves knowing something about it' (351). The crucial point, again, is whether this 'something' which we know about the concept comes to us via our intuitions (arguably also *somehow* informed by our experience and knowledge about the world) or whether

---

[11] Schupbach invites his readers to look to *experimental philosophy* to establish similarity between the explicatum and the explicandum.

we develop it on the basis of a close study of scientific practice. It is the latter approach which I advocate as a methodological framework for part of the discussion in this book.

Recall our discussion of the notion of ad hoc hypotheses in Chapter 5. The meaning of ad hoc hypotheses seems intuitive indeed: an ad hoc hypothesis is a hypothesis that was introduced in order to save a theory from refutation. And yet it is deficient in that it does not tell us what it is about these hypotheses that is epistemically objectionable – it only tells us about motivations. Furthermore, the use of ad hoc hypotheses in scientific practice is complex. For example, it exhibits a time dimension which some have taken to be evidence for its subjectivity. And although scientists are capable of making ad hoc judgements in the right conditions, they hardly ever reflect on these conditions and presumably are also not exempt from making erroneous ad hoc judgements. Furthermore, scientists may not explicitly use the term 'ad hoc', even when that is what they mean. There is thus plenty of work that offers itself to the philosopher of science interested in concept clarification.

In our explication of ad hocness, we saw that various previous attempts are descriptively inadequate (such as the independent support account) or conceptually problematic (such as the unificationist account). Instead, we proposed that ad hoc judgements are judgements about theoretical coherence, understood in a specific way. Carnap's desideratum of similarity we met by a close study of how the concept is used in scientific practice. Carnap's desideratum of exactness and simplicity we met by providing clear identity conditions (that the account is simple can easily be gleaned from a comparison to Leplin's list of conditions). The coherence account of ad hocness is also fruitful in that it allows us to clarify various episodes in the history of science in which the concept of ad hoc hypotheses played a crucial role, without having to surrender to subjectivism.[12] Studying the history of science is perhaps not necessary for the kind of concept clarification advertised here; maybe immersing oneself in contemporary scientific practice will suffice.[13] However, studying the history of science clearly

---

[12] Another concept which I have sought to clarify along these lines elsewhere is that of scientific discovery; see Schindler (2015). The notion of novel success (Chapter 3), I think, is yet another example of historically informed concept clarification.

[13] Giere (1973) argued that there is no need to consult the history of science, as an analysis of contemporary science suffices for the analysis of science, and mentioned as examples quantum mechanics, molecular biology, and contemporary psychology. Laudan (1977, 159) points out that 'the fact that a scientific theory is still believed and is currently undergoing development scarcely makes that theory ahistorical . . . Giere's own historical preferences may be for the recent past, but they are nonetheless historical for all that'.

constitutes one way in which clarification of concepts central to science can be carried out in a fruitful way.

Finally, let us also note that the project of clarifying scientific concepts in view of the use of these concepts in scientific practice, as revealed by historical studies, runs no obvious risk of violating the norm–fact divide. This project is primarily about how scientists *actually* use certain concepts. The project is therefore primarily descriptive. However, it is not *purely* descriptive, because, as we have already noted, scientists hardly ever reflect on the conditions of satisfaction of the concepts they use; those conditions need to be reconstructed. But because scientists do not know these conditions, they cannot by themselves arbitrate on the basis of these conditions any disputes concerning the use of concepts. For that, they need the help of philosophers. An explicatum that meets all of Carnap's desiderata (although admittedly difficult to achieve) might thus have real practical value. First and foremost, however, concept explication will allow us to develop a better philosophical understanding of science, with both descriptive and normative force.

Conclusion

This book revisited the long-standing question What are the good-making features of a scientific theory, or virtues, for short? It was the goal of the book to provide an adequate assessment of theoretical virtues in order to then spell out the appropriate consequences for realism. The results are as follows.

It has often been denied that simplicity is an epistemic theoretical virtue, because 'we don't know whether the world is actually simple'. However, this is a misapprehension of what is required for simplicity to be an epistemic virtue; the world need not be simple (whatever that is supposed to mean) for simplicity to be an epistemic virtue. Instead, what is required is only that it be shown that simplicity makes a theory more likely to be true.

In this book, I offered an *evidential-explanatory rationale* for the epistemicity of simplicity, which reduces simplicity considerations in theory choice to evidential-explanatory ones. More specifically, the evidential-explanatory rationale demands that one adopt only theories that postulate entities or principles that are empirically supported in the explanation of the phenomena. Entities or principles which are not required for explaining the phenomena are explanatorily superfluous and thus not empirically supported. Hence, theories that are simple in that they postulate no superfluous entities or principles are to be preferred to theories that do postulate such superfluous entities or principles, simply because the former postulate no entities or principles that are not empirically supported in the explanation of the phenomena.

The evidential-explanatory rationale for simplicity is context-dependent in two ways. First, what *form* of simplicity (syntactic or ontological) is relevant in comparing theories, for example, depends on the context in which the theory choice is made and is to be determined by the scientific community in question. Second, the evidential-explanatory rationale does not warrant any *absolute* preference for simplicity. It licenses only a preference amongst those theories which explain the phenomena. Thus, a theory T_1 that postulates 61 fundamental particles (as the standard model in particle physics does) is preferable to a theory T_2 that postulates 78 fundamental particles and explains the same phenomena, because T_2 postulates more particles (namely exactly 17 more) than are required for explaining the same phenomena. Yet, a theory T_3 that postulates only 32 fundamental particles, despite being simpler than T_1, would only be preferable to T_1 if those 29 additional particles postulated by T_1 were not needed for explaining the phenomena. For the antirealist, two empirically equivalent theories warrant the same belief, no matter their degree of complexity. But according to the evidential-explanatory rationale, this is not the case: a theory that explains the same facts with fewer principles or entities is indeed better supported by those facts than more complex theories. Belief in simpler theories is thus warranted. Contrary to the antirealist, simplicity is indeed an epistemic virtue of theories. This was my *first argument for realism*, namely the *argument from simplicity*.

My *second and central virtuous argument for realism*, namely the *no-virtue-coincidence argument*, showed that a 'very virtuous' theory – i.e., a theory that possesses all of the five standard virtues – is likely to be true. I argued this on the basis of two elements: (i) the Kuhnian framework for theory choice in which scientists, despite their divergent theory-choice preferences, can agree on the same theory when the theory possesses all the relevant virtues; and (ii) Earman's observation that the probability of an event having occurred converges to 1, even when the base rate tends to 0, so long as the number of witnesses having observed the event converges to infinity. Analogously, the more scientists embrace a theory, the more likely it is that the theory is true, even when the base rate of true theories is diminishingly small. Although this is no blanket victory for the realist (the number of scientists is limited at any given time), it does show, contrary to Magnus and Callender, that a rational debate about realism can be had: with the no-virtue-coincidence argument (NVC) in hand, realists and antirealists can focus their efforts on arguing about the error rates, which – in contrast to the base rates – are very much accessible. For the realist to succeed, the virtues need not be absolutely truth-conducive, but only *relatively* truth-conducive. More specifically, the true positive rate of virtuous and true theories need not be larger than 0.5, but only larger than the false positive rate of virtuous but false theories. Furthermore, the inverse relationships between the true positive and the false negative rates, on the one hand, and the false positive and true negative rates, on the other, set important constraints for the values each of these rates can take. I have argued that within these constraints, and with the appropriate charity towards the antirealist, a good case can be made that some theoretical virtues are indeed relatively truth-conducive.

One theory property which is almost universally recognized as a theoretical virtue is 'non–ad hocness'. Common rationales for predictivism – i.e., the view that novel success has more evidential weight than accommodated evidence – are also motivated via this notion. Yet, curiously, there has been very little discussion of it in the past few decades. In this book, I revisited the question of ad hocness and argued for a coherentist conception of ad hocness. According to that notion, a hypothesis H is ad hoc, when it does not cohere with any background theories B or the theory T it saves from refutation, whereby coherence is to be understood as the provision of good theoretical reasons by T or B for belief in H. My account of ad hocness supports my *third virtuous argument for realism*, that is, the *argument from coherence*: a theory's coherence is a sign of its truth, because a theory can provide good reasons for believing in the hypotheses it invokes to accommodate the phenomena when the same structure underlies these phenomena, whereas it cannot provide such reasons when such a structure is absent. Although the discovery of real structures underlying the phenomena is not guaranteed through coherent theories, in a well-functioning science, the discovery of such principles should at least be more likely with coherent theories.

Empiricists like van Fraassen have argued forcefully that theoretical virtues are only pragmatic virtues, not epistemic ones. That is, although empiricists grant that theoretical virtues such as simplicity and unifying power do have important roles to play in theory choice, they only concern the convenient use of theories; they tell us nothing about whether a theory is likely to be true. This Negative View about theoretical virtues, as I dubbed it, entails what I called the 'dictatorship condition', according to which the empirical adequacy of a theory always overrules concerns about virtues such as simplicity and unifying power in theory-choice situations.

I argued in this book that the dictatorship condition has been violated in some of our most stunning scientific discoveries, such as light-bending, the structure of DNA, the periodic table, special relativity, plate tectonics, and the weak neutral current. In all of these cases, the scientists making the discoveries chose to set aside evidence that apparently contradicted their relevant theories and decided to adopt those theories on the basis of evidence that supported those theories. I argued that we know that they adopted and not just pursued those theories because, in fact, they actively dismissed the negative evidence that others took very seriously. I suggested that the theoretical virtues of the theories in question functioned as confidence boosters: the virtues gave scientists reasons for belief that their theories were true *despite* the apparently negative evidence. I argued further that we must accept that those reasons are *good* reasons and, accordingly, that the Negative View is false, lest we want to render these groundbreaking discoveries the results of utterly irrational theory choice decisions (which we shouldn't). This was my *fourth virtuous argument for realism*, that is, the *argument from choice*.

Although my four virtuous arguments for realism are independent, the arguments from simplicity, coherence, and choice all support my no-virtue-coincidence argument, NVC, which feeds off the view that theoretical virtues are relatively truth-conducive. The argument from simplicity strengthens the NVC in that it shows that one of the standard virtues it appeals to is, in fact, an epistemic virtue. Although coherence did not explicitly figure in our presentation of the NVC, there are good reasons to include it (a theory ought not to be ad hoc, *mutatis mutandis*). The argument from choice provides more general support from scientific practice for the view that theoretical virtues raise the probability of the truth of a theory.

The realist position presented in this book, *virtuous realism*, differs from standard accounts in that it does without the usual emphasis on novel success. Standard accounts tend to make their realist commitments conditional on whether a theory manages to produce novel success. Realists, for example, allow only those theories to figure in the base of the pessimistic meta-induction which managed to produce novel success. Realists have also singled out novel-success-producing theory parts when applying the divide et impera move to false but successful theories. Whether these arguments for realism are compelling, therefore, critically depends on whether it is plausible that novel success is better evidence than accommodative success – a view also known as predictivism. Often this is just assumed to be the case in the realism debate. However, as I've argued in this book, none of the rationales proposed for why predictivism ought to be true is compelling. Virtuous realism, in contrast, is not vulnerable to such criticisms, as none of my virtuous arguments appeal to predictivism.

There is a much weaker view of fertility than what predictivism would demand: a theory is fertile if it is able to accommodate anomalies in a non–ad hoc fashion by means of *de-idealization* of its original simplifying assumptions. McMullin believes that this view of fertility in fact supports realism better than predictivism. But here I am also sceptical. Although I think McMullin has correctly identified an important form of theoretical fertility that is virtuous, I do not think it is appropriately captured by the notion of de-idealization; it therefore does not support McMullin's argument for realism either. Nevertheless, a theory's capacity to accommodate anomalies in a non–ad hoc fashion *can* support realism via my *argument from coherence*.

The argument from coherence may in fact open a new way of thinking about progress in science. Instead of seeking to continue to engage with antirealists in debates about which parts of our past false theories might have been responsible for generating novel success, the realist may want to direct her attention to the ways in which theories explain the phenomena. When comparing past and present theories, it is noteworthy that theories that superseded earlier theories seem to explain the phenomena in a less ad hoc way. The Copernican system, for example, explained many of the planetary phenomena from first principles, which in the Ptolemaic system had to be saved with ad hoc assumptions. Whereas there was no theoretical reason for Newton to assume any particular form of his inverse square law (why square rather than, for example, cube?), the gravitational force in Einstein's theory must have the form that it has (Weinberg 1993, 108). Even in theories scientists are working towards, we can see attempts to reduce the number of ad hoc assumptions, as in the development of string theory, where theorists have sought to give theoretical reasons for assuming particular values for the plenitude of free parameters of the standard model. Overall, then, there appears to be a progression towards theories with fewer and fewer ad hoc assumptions, or, as Weinberg would say, towards ever greater inevitability in our theories. If my argument from coherence is sound, the decrease in ad hocness (and increase in coherence) tells us that our theories are getting closer and closer to the truth – even when they will never in fact reach it, just as we shouldn't expect to ever possess a theory without any ad hoc assumptions.

Whether or not virtuous realism will prove to be a robust alternative to standard accounts of realism remains to be seen. At the very least, I hope to have inspired the reader to reconsider the ways in which we normally think about theory choice in science and what we normally take to be our best reasons for believing what our theories tell us about the world.

Epilogue: The Demarcation Problem

At the beginning of this book, we noted that there is a strong particularist trend in contemporary philosophy of science away from the more global questions about science towards more local, discipline-specific questions. I sought to resist this trend and to reinvigorate some of the classic methodological debates in the philosophy of science which philosophers have lost sight of recently. With this book, questions such as the following should now have a fresh basis for discussion: What are the good-making features of a scientific theory? What does it mean for a scientific theory to be ad hoc? What role do a theory's virtues play in actual theory choice? What kind of evidence matters most in the confirmation of a scientific theory? A question we have not yet addressed is what it may be that makes the theoretical virtues we have discussed virtues of good *scientific* theories. Answering this question would require that we address the question of what distinguishes science from non-science or pseudoscience. This is the venerable demarcation problem. I want to use the remaining pages to bring that problem into sharper focus.

The problem of demarcation can be understood broadly as the demarcation of science against all intellectual enterprises which are not science – for example, philosophy, religion, logic, mathematics, and engineering. Or it could be understood more narrowly as the problem of demarcating science from pseudoscience, i.e., intellectual enterprises which see themselves as competing with science with regard to knowledge claims but which in fact fall far behind the methodological standards of science. Often-cited pseudosciences are astrology, alchemy, alternative medicine, intelligent design (i.e., creationism), and psychoanalysis.

The logical positivists were primarily interested in the broader question and, in particular, in demarcating science from metaphysics, whereby their criterion of verifiability – i.e., whether a statement could be related to experience or even perception – was supposed to determine whether a statement should be viewed as meaningful or meaningless. As Popper once aptly put it, in logical positivism, 'verifiability, meaningfulness, and scientific character all coincide' (Popper 1963/1978, 40). Although Popper was not nearly as dismissive about metaphysics as were the logical positivists, Popper himself conceived of the demarcation problem in a broad way initially, but later gave it the more narrow reading (Popper 1959a, 1963/1978). Let us in what follows understand the demarcation problem in this more narrow sense, unless otherwise indicated.

The demarcation problem, as Lakatos once put it, 'is not a pseudoproblem of armchair philosophers', since it has important societal ramifications (Lakatos 1978, 1). It is, for example, capable of affecting public and legal disputes, research prioritizations, funding decisions, and education. For example, is homeopathy or Chinese medicine a legitimate form of medical cure? Is cold-fusion research worthy of funding? Is creationism a legitimate competitor to the theory of evolution? Should we teach creationism at school?[1] Because the stakes are so high, any proposed alleged solution to the demarcation problem should therefore be as precise as possible and 'especially compelling' (Laudan 1983, 120).

By far the most famous demarcation criterion is Popper's criterion of falsifiability or testability (which I will use interchangeably here). It turns out, however, that it is a rather weak criterion. As some have argued, a clear pseudoscience such as 'intelligent design' makes falsifiable predictions, although it is easy to show that some of these predictions are false (e.g., the prediction that animal and human fossils ought to be found in not-too-distinct geological strata, as God supposedly created humans and animals on the same day) (Laudan 1982). Popper himself recognized that there were theories, such Marx's historical materialism, which he was inclined to classify as pseudoscience even though they were falsifiable. What he objected to in such cases were not so much the theories themselves, but rather the manoeuvres by the theories' proponents to 'escape refutation' – 'for example, by introducing ad hoc some auxiliary assumption, or by re-interpreting the theory ad hoc' (Popper 1966, 37). In contrast, psychoanalysis, for Popper, was 'in a different class'; it was 'simply non-testable, irrefutable' (37).[2] Yet, even this has been debated. Grünbaum, in a string of

---

[1] There are several court cases in the United States where it had to be decided whether creationism should be taught at school. One of them, *McLean v. Arkansas Board of Education* (1981/2), gained special prominence amongst philosophers because of the involvement of philosopher Michael Ruse (as expert witness) and philosopher Larry Laudan's critical reply to the ruling against creationism (Laudan 1982).

[2] One of Popper's two main examples to support this claim is the one of a man drowning a child and a man saving a child from drowning (cf. Chapter 1).

publications, has argued that Popper's examples were 'contrived' and that psychoanalysis as *actually* embraced by Freud and others did have falsifiable consequences and, furthermore, that psychoanalysts were very much pre-pared to give up their theories when confronted with contrary evidence (Grünbaum 1979).

Another pseudoscience explicitly identified by Popper is astrology. However, it too slips through the hands of the falsifiability criterion. In a paper appearing in *Nature* in 1985, it was shown in meticulous detail in a double-blind test that the 'fundamental thesis of natal astrology', namely the idea that the positions of heavenly objects at one's date of birth determine one's character and destiny, could be tested (and proven false) (Carlson 1985). Kuhn (1970a), in his perceptive criticism of Popper's demarcation criterion, also noted the falsifiability of historical astrology and furthermore argued that its practitioners could not be faulted for explaining their predictive failures in terms of the complexity of the causal factors at play, since today's meteorologists and medical researchers do so too (legitimately) (8).

The above examples indicate that testability is not a sufficient condition for scientificity (if it was ever meant to be). Is it even a necessary condition? It is well known that Popper once believed that the theory of evolution was not falsifiable, and therefore not scientific, although he did believe that it was effective in giving rise to testable models (Popper 1976; cf. Section 1.2.3).[3] But clearly, evolution itself is as good a scientific theory as any. To the extent that evolution is not testable (which is of course disputed; see Ruse 1977), testability cannot be a good demarcation criter-ion. Interestingly, there is a scientific theory that is indisputably untestable as it currently stands, namely string theory. Originally developed as a theory of strong interactions in the 1970s, it soon developed into a 'theory of everything', i.e., a theory of all four forces of nature. String theory is based on the idea that the fundamental physical entities in the world are not particles but strings and membranes instead (higher-dimensional extensions of one-dimensional strings). Even after decades of development, string theory still faces a plethora of theoretical problems

---

[3] Popper later changed his mind, but for rather strange reasons (Popper 1978). He reasoned that the theory of evolution is 'not only testable, but it turns out to be not strictly universally true', because some biological traits are caused by genetic drift rather than natural selection (Popper 1978, 346). In fact, the alleged falsity of evolution in these cases seems to have been Popper's only reason for deeming the theory testable. A more reasonable assessment would probably be that – like all theories – the theory of evolution has a certain scope within which it is true. It is not false in instances that lie outside its scope.

(e.g., the unintuitive postulation of six extra dimensions in addition to spacetime, the incredibly high number of vacuum states, and the lack of full coherency), but it is currently without alternative when it comes to the unification of all four forces of nature (Dawid 2013).

Although string theory is consistent with the physics described by quantum mechanics and the general theory of relativity, it has thus far not generated any predictions of its own. This has been criticized and led to Popperian claims that string theory is not a science or perhaps better given up in favour of testable approaches (Cartwright and Frigg 2007; Smolin 2007; Woit 2011; Ellis and Silk 2014). Yet, one can appreciate that testability is a desideratum in science and that practitioners themselves in fact strive for it, without having to conclude that the lack of testability makes string theory end up on the wrong side of the demarcation divide (Johansson and Matsubara 2011; Camilleri and Riston 2015). On the contrary, given that string theory is pursued by a large part of the *scientific* community of theoretical high-energy physicists, any good naturalist should conclude that testability is too strong a demarcation criterion – particularly when the same criterion entails that creationism and astrology should be deemed sciences!

Even if testability were a good demarcation criterion for scientific *theories*, it would not be enough to demarcate science with its theories and *practices* from pseudoscience. As we have pointed out already in Chapter 1, Popper himself conceded that an empirical test never speaks unequivocally against the theory under consideration; it might just indicate a flaw in one of the auxiliary assumptions we make (consciously or not) when we conduct an empirical test. Since an empirical test cannot logically force us to refute a hypothesis after a negative test, then, so Popper thought, practitioners should be prepared to seek the error in the tested theory itself, not in the auxiliaries. In other words, practitioners should be willing to falsify their theories.[4]

Kuhn, in contrast to Popper, emphasized that for science to advance, scientists must *accept* a set of common assumptions and practices, namely a paradigm. More specifically, committing to a paradigm will allow scientists to engage with involved conceptual and empirical problems, which arise from this commitment, in a highly efficient and effective way. Kuhn likened such problems to puzzles, because he was convinced that paradigm problems, like puzzles, gave scientists the expectation of the availability of

---

[4] Many of the contributions to a recent volume on the demarcation problem focus on the practices (rather than the theories) of pseudoscientists (Pigliucci and Boudry 2013).

a solution. He believed that this expectation was so strong that a failure to solve problems would normally result in blame being put on the scientist rather than on the theory in question. Although the idea of paradigm and puzzle-solving was foremost meant as a description of how science actually proceeds, it was also supposed to demarcate science from pseudoscience (Kuhn 1970a).[5]

In an exchange with Kuhn, Popper retorted that Kuhn's ideas had no place in 'pure' science, but only in 'applied' science, where one 'learned a technique which can be applied without asking for the reason why' and where the main problem the scientist is concerned with is 'a routine problem, a problem of applying what one has learned'. Popper equated the acceptance of a paradigm with being uncritical, which he thought was 'a danger to science and, indeed, to our civilization' (all quotations from Popper 1970, 53). Nevertheless, in apparent contradiction to much else that he said, Popper did concede that 'some dogmatism' was necessary, as otherwise 'we shall never find out where the real power of our theories lies' (55). Lakatos's account of research programmes and their hard cores 'protected from refutations' is built on this Kuhnian insight as well (Lakatos 1978).

Thus far, we have assumed that a successful solution to the demarcation problem should specify necessary and/or sufficient conditions of scientificity. But need that be so? Laudan (1983) argued yes. If a demarcation criterion provides only necessary conditions N for scientificity, then we can say on the basis of N that, e.g., astrology is not a science when it does not satisfy N, but we cannot say whether, e.g., physics is a science when it *does* satisfy N. On the other hand, if we are given only sufficient conditions S, then we can say that physics is a science when it meets S, but we cannot say that astrology is not scientific when it does not satisfy S. Because Laudan did not see any plausible contenders for solving the demarcation problem in terms of necessary and sufficient conditions, he concluded that the demarcation problem was 'spurious' and a 'pseudo-problem' (124). We should therefore dispense with terms like 'unscientific' entirely (125).

Rather than trying to distinguish between science and non-science, Laudan suggested that we simply weed out the bad theories – whether scientific or not – by subjecting them to empirical tests (125). Such a deflationist strategy, however, cannot be fully satisfactory, as it would not allow us to draw the line that deserves to be drawn between creationism and evolution: these two theories simply do not have the same scientific

---

[5] See Chapter 7 and Schindler (2014b) for a discussion of normativity in Kuhn's account.

status. Creationism fares worse than evolution not only because it is a theory that makes more false claims about the world than the theory of evolution. That would also be true for classical Mendelianism as compared with molecular genetics. No, there seem to be elements in creationism that have no place in a scientific theory. For example, no serious contender for a scientific theory should postulate supernatural powers for solving the problems in its domain. But Laudan's deflationist strategy would not allow us to raise such issues; it would restrain our focus merely on empirical testing. Such a restriction seems particularly problematic in the educational context: we should not want to grant creationism the appearance of a legitimate (but false) scientific theory. In such contexts, Laudan's deflationism lacks the normative force of accounts that do not try to resolve the demarcation problem.

Not only is Laudan's deflationism problematic; his reasons for rejecting the demarcation problem are fallacious too. It is true that without necessary and sufficient conditions we cannot say *with logical certainty* whether something is not a science or is a science. However, I don't think we need to set such high demands on a successful solution to the demarcation problem. Instead, we might just require, as I indicated earlier with regard to theoretical virtues, that something *should* (but need not necessarily) follow certain criteria of scientificity N and that it *probably* would not qualify as a science if it were not to satisfy N. Likewise, we would say that something *probably* is scientific when it conforms to criteria S. There are two interpretations of this way of speaking. One would be that the demarcation criteria are categorical (e.g., either testable or not), and the probabilities only express our limited cognitive abilities in determining whether or not the intellectual enterprise in question satisfies these criteria. Another interpretation would be that the demarcation criteria are not categorical, but rather allow for degrees of scientificity. Pseudosciences could therefore be 'highly dogmatic', for example, whereas scientific ideas (embedded in a paradigm) would not be as dogmatic (but still to some degree). There would thus be no clear boundary between science and non-science, but only a spectrum of scientificity. Such a view of science has recently been defended by Hoyningen-Huene (2013), who holds that science is not really distinct from common sense, but only more systematic.

A view compatible with such way of speaking is a view that has gained a lot of popularity recently amongst philosophers, according to which the class of all sciences is best thought of in terms of the Wittgensteinian idea of *family resemblance* (Dupré 1993; Irzik and Nola 2011; Hoyningen-Huene 2013; Pigliucci 2013). According to Wittgenstein (1953), many of the general terms or kind terms which we use in our language denote sets of entities which are only loosely bound together by similarity; there is nothing that they all have in common. Accordingly, there cannot be any necessary or even sufficient conditions for what the general term denotes or for the kind membership. Wittgenstein's example concerns the term 'game', which we use for a plenitude of activities, such as card games, board games, ball games, computer games, schoolyard games, hide-and-seek, and even mind games. Wittgenstein concluded that all we can say about such sets of entities denoted by the same general term is that there is a 'complicated network of similarities overlapping and criss-crossing'. In the case of science, one would say that although there are similarities between the various sciences, there is no cut-off point at which we could say an activity that does not satisfy condition N cannot be a science. Likewise, we cannot say anything like that for activities within science; i.e., we cannot say more than that the sciences resemble each other in some ways.

There is a famous problem with the concept of family resemblance, known as the problem of 'wide-open texture' and first raised by Richman (1962). If, as the idea of family resemblance has it, things denoted by a kind term are individuated on the basis of similarity, and since, as Goodman (1972) pointed out, 'anything is similar to anything else', then on what basis can we '*refuse* to apply the term to anything'? Boxing and street fighting are similar in many ways (in both activities, fists are thrown in order to harm the opponent), yet only one of them is a sport (Pompa 1967). So if it is similarity that binds together sets of things in family resemblance, on what basis would we refuse to subsume street fighting under the term 'sport'?

The problem does not only concern the negative delimitation of kinds; it also concerns their positive delimitation. My brother and I are similar: we are both tall and have protruding ears. Does that make us members of the same family? No. My brother and I are members of the same family in virtue of our hereditary relations. Likewise, there are many similarities between the sciences which are not suitable for justifying membership in the science kind. For example, physics and chemistry are similar in that most of their practitioners are white men. Yet we would rightly consider this similarity as outright irrelevant to the question of what it is that makes them members in the science kind. There has got to be something else that picks out the *right* similarities for us. Arguably, it is the kind identity itself that determines which similarity relations are relevant to the individuation of the kind and which ones aren't (cf. Pompa 1967). That is, in the case of biological families, which are defined in terms of hereditary relations, it is those hereditary relations which determine that a resemblance (e.g., my and my brother's protruding ears) is a resemblance within a particular family rather than between members of different families (e.g., my and Barack Obama's protruding ears). In other words, it seems we must define the kind identity *before* we can single out certain similarities that can justify kind membership. But then, what is the work that family resemblance does for us? And what normative force would such a solution, which gives such meagre grounds for in- and exclusion, have in any case?

Simon (1969) insists, contrary to Wittgenstein, that 'when we do use a word to apply to things that share no common essential feature, we are using it in different senses' (412). In other words, Simon considers it a necessary condition that a set of entities have something in common for us to be justified in using a general term with the *same meaning* for that set. That is, the meaning of the term 'game' would be a different one when we apply it to such different activities as chess and schoolyard ball games (the former involves a strict set of rules, the latter doesn't). But, of course, those different meanings of the word game must somehow be related in order for them to be recognizable as being subsumed by the same term (Simon also speaks of a 'family of meanings').

Simon proposes that there is a paradigm example of each general term or kind term which establishes commonality of different meanings of a single such term.[6] Say there is an activity *a* which is a G1 and an activity *b* which is a G2, and *a* and *b* share no common features; then *a* and *b* can still be recognized as instances of G (which includes both G1 and G2) because there is a paradigm example of G, namely *g*, which has a feature in common with both *a* and *b*. The features of the paradigm *g* therefore delineate the set of all activities that can be a G. However, the features of *g* do not set out necessary and sufficient conditions for all members of G, since there clearly are members, such as *a* and *b*, which do not share any features (see Figure E.1).

In the case of science, Simon's paradigm idea could translate as follows. Suppose activity *a* is an instance of biology, activity *b* is an instance of psychology, and activity *c* is an instance of chemistry. Even though biology, chemistry, and psychology are very different, and even though there might not even be one thing they all share, they would still be recognizable as science because they all have at least one property they share with the

---

[6] I take it that Simon's view of a kind-defining paradigm has affinities to Eleanor Rosch's famous psychological theory of human categorization, namely 'prototype theory'.

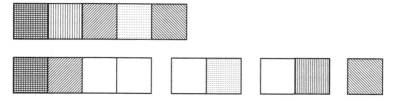

Figure E.1 The Science Paradigm view. The upper set of boxes represents a kind paradigm or basic predicate. The lower sets of boxes represent instances of the kind which either have very little or even nothing in common and nevertheless each share at least one feature with the paradigm/basic predicate.

paradigm science (e.g., physics) or with the 'basic predicate' (as we shall see in a moment). As Simon puts it,

> [w]e do not in fact insist that a discipline cannot properly be called a science, or a procedure be called scientific, if it lacks any one of the features of being experimental, mathematical, systematic, and painstaking; and yet we demand all of these characteristics (and perhaps other besides) of a paradigm, such as physics. (Simon 1969)

For Simon, the paradigm does not actually have to exist; it may just be 'conceived'. This proposal, made already by Pompa (1967), is known as the 'basic predicate solution' (cf. Bellaimey 1990). We may thus refer to this solution as the paradigm/basic predicate solution or the PBP solution, for short. An advantage of this solution is that it is not tied to any particular science: even our best example of science might not incorporate all the features that we would want to associate with science.

It is worth noticing that even though there are no necessary and sufficient conditions for a kind membership according to the PBP solution, there is still a set of properties that is necessary and sufficient for the PBP *itself.* Since the PBP individuates the kind, the solution may thus be called an 'essentialism in disguise'. Accordingly, a change in the PBP will also change the set of instances that are picked out by it, which in turn should amount to a change in the meaning of the relevant kind term. This is not too implausible, I think. When zoologists decided to incorporate the taxon monotremes into the natural kind mammal, the meaning of the latter arguably changed, as it now includes a feature which was ruled non-mammalian before that – namely, the feature of egg-laying. The same is true for zoologists' decision to exclude whales from the kind fish (Kuhn 1990; LaPorte 2009). In the case of science, there have been radical inventions, such as use of statistics in experimental tests in the late nineteenth century, which, one might argue, have altered what we mean when we use the term 'science'.

It is time for us to take stock now. A central attraction of the family resemblance concept is that it allows for 'fuzzy' boundaries: there are no criteria that would categorically rule some activity in or out of a term's extension. This seems appropriate for the term 'science', as science seems too diverse for it to be plausible that there could be any necessary and sufficient criteria for scientificity that would capture *all* activities which we regard as scientific. However, as we just saw, the family resemblance concept is beset with problems. Not only is there the well-known problem of wide-open texture, but it is also questionable whether similarities can really *justify* kind membership without some prior relevance determination through the kind in question. But without such justifications, it is hard to see how a solution to the demarcation problem based on the concept of family resemblance could have any normative force.

The PBP solution, as we have seen, offers something of a compromise between family resemblance and essentialism: although, just like on the family resemblance solution, there is not one feature that all sciences must share, there *are* necessary and sufficient conditions for the PBP, which delineates the kind. More specifically, this means that it is necessary that a science possess at least one feature that characterizes the PBP. And possession of *at least* one of the features of the PBP is sufficient for an activity to be a science.

Whether or not we'll ever manage to identify the right demarcation criteria is an open question. Indeed, I have said nothing here about how this might be done practically. My goal in this epilogue was more basic: to argue that the sciences must have something in common for us to *reliably* and *correctly* apply the term 'science' to physics, chemistry, and biology, and reliably and correctly deny application of the term to homeopathy, astrology, and creationism. Nevertheless, my view is that the right demarcation criteria for scientific theories will coincide with the criteria for *good* scientific theories, namely theoretical virtues.

Bibliography

Achinstein, P. 1974. History and Philosophy of Science: A Reply to Cohen. In *The Structure of Scientific Theories*, F. Suppe (ed.), Urbana: University of Illinois Press, 350–360.

1994. Explanation v. Prediction: Which Carries More Weight? *PSA: Proceedings of the Biennial Meeting of the Philosophy of Science Association*, Vol. 2: *Symposia and Invited Papers*, 156–164.

Acuña, P. 2014. On the Empirical Equivalence between Special Relativity and Lorentz's Ether Theory. *Studies in History and Philosophy of Science Part B: Studies in History and Philosophy of Modern Physics*, **46** (May): 283–302.

Aubert, J. J., U. Becker, P. J. Biggs, et al. 1974. Experimental Observation of a Heavy Particle J. *Physical Review Letters*, **33** (23): 1404–1406.

Augustin, J. E., A. M. Boyarski, M. Breidenbach, et al. 1974. Discovery of a Narrow Resonance in e+e– Annihilation. *Physical Review Letters*, **33** (23): 1406–1408.

Baker, A. 2003. Quantitative Parsimony and Explanatory Power. *British Journal for the Philosophy of Science*, **54** (2): 245–259.

2013. Simplicity. *Stanford Encyclopedia of Philosophy* (fall 2013 edition), E. N. Zalta (ed.), http://plato.stanford.edu/archives/fall2013/entries/simplicity/.

Bamford, G. 1993. Popper's Explications of Ad Hocness: Circularity, Empirical Content, and Scientific Practice. *British Journal for the Philosophy of Science*, **44** (2): 335–355.

1996. Popper and His Commentators on the Discovery of Neptune: A Close Shave for the Law of Gravitation? *Studies in History and Philosophy of Science Part A*, **27** (2): 207–232.

Barnes, E. C. 1992. Explanatory Unification and the Problem of Asymmetry. *Philosophy of Science*, **59** (4): 558–571.

2000. Ockham's Razor and the Anti-Superfluity Principle. *Erkenntnis*, **53** (3): 353–374.

2005. On Mendeleev's Predictions: Comment on Scerri and Worrall. *Studies in History and Philosophy of Science Part A*, **36** (4): 801–812.

2008. *The Paradox of Predictivism*. Cambridge: Cambridge University Press.

Baron, S. and J. Tallant. in press. Do Not Revise Ockham's Razor without Necessity. *Philosophy and Phenomenological Research*, https://onlinelibrary.wiley.com/doi/full/10.1111/phpr.12337.

Bellaimey, J. E. 1990. Family Resemblances and the Problem of the Under-Determination of Extension. *Philosophical Investigations*, **13** (1): 31–43.

Bernabeu, J. and C. Jarlskog. 1977. Relations among Neutral Current Couplings to Test the SU (2)⊗ U (1) Gauge Group Structure. *Physics Letters B*, **69** (1): 71–76.

Bjorken, J. D. 1972. Theories of Weak and Electromagnetic Interactions Employing the Higgs Phenomenon. In *16th International Conference on High-Energy Physics, Batavia, Illinois, 6–13 Sept. 1972*, J. D. Bjorken, J. D. Jackson, A. Roberts, and R. Donaldson (eds.).

1977. Alternatives to Gauge Theories. In *Unification of Elementary Forces and Gauge Theories*, D. B. Cline and F. E. Mills (eds.), London: Harwood Academic Publishers, 701–726.

1979. Neutral-Current Results without Gauge Theories. *Physical Review D*, **19** (1): 335.

Bjorken, J. D. and C. H. Llewellyn Smith. 1973. Spontaneously Broken Gauge Theories of Weak Interactions and Heavy Leptons. *Physical Review D*, **7** (3): 887–902.

Bogen, J. and J. Woodward. 1988. Saving the Phenomena. *Philosophical Review*, **97** (3): 303–352.

Bohr, N. 1915. On the Series Spectrum of Hydrogen and the Structure of the Atom. *Philosophical Magazine*, **29**: 332–335.

Bondi, H. 1960/1952. *Cosmology*. London: Cambridge University Press.

Bondi, H. and T. Gold. 1948. The Steady-State Theory of the Expanding Universe. *Monthly Notices of the Royal Astronomical Society*, **108**: 252.

BonJour, L. 1985. *The Structure of Empirical Knowledge*. Cambridge: Cambridge University Press.

Boudry, M. and B. Leuridan. 2011. Where the Design Argument Goes Wrong: Auxiliary Assumptions and Unification. *Philosophy of Science*, **78** (4): 558–578.

Boyd, R. N. 1983. On the Current Status of the Issue of Scientific Realism. *Erkenntnis*, **19** (1–3): 45–90.

Brössel, P., A.-M. A. Eder, and F. Huber. 2013. Evidential Support and Instrumental Rationality. *Philosophy and Phenomenological Research*, **87** (2): 279–300.

Brown, M. J. and I. J. Kidd. 2016. Introduction: Reappraising Paul Feyerabend. *Studies in History and Philosophy of Science Part A*, **57** (June): 1–8.

Brush, S. G. 1976. *The Kind of Motion We Call Heat: A History of the Kinetic Theory of Gases in the 19th Century*. Amsterdam: North-Holland Publishing Company.

1989. Prediction and Theory Evaluation: The Case of Light Bending. *Science*, **246** (4934): 1124.

1994. Dynamics of Theory Change: The Role of Predictions. *PSA: Proceedings of the Biennial Meeting of the Philosophy of Science Association*, Vol. 2: *Symposia and Invited Papers*, 133–145.

1996. The Reception of Mendeleev's Periodic Law in America and Britain. *Isis*, **87** (4): 595–628.

Burian, R. M. 1990. Review: Andrew Pickering, Constructing Quarks. *Synthese*, **82**: 163–174.

Camilleri, K. and S. Ritson. 2015. The Role of Heuristic Appraisal in Conflicting Assessments of String Theory. *Studies in History and Philosophy of Science Part B: Studies in History and Philosophy of Modern Physics*, 51: 44–56.

Carlson, S. 1985. A Double-Blind Test of Astrology. *Nature*, 318 (6045): 419–425.

Carman, C. and J. Díez. 2015. Did Ptolemy Make Novel Predictions? Launching Ptolemaic Astronomy into the Scientific Realism Debate. *Studies in History and Philosophy of Science Part A*, **52**: 20–34.

Carnap, R. 1937. *The Logical Syntax of Language*. London: K. Paul, Trench, Trubner & Co.

1950. *Logical Foundations of Probability*. Chicago, IL: Chicago University Press.

Carrier, M. 1991. What Is Wrong with the Miracle Argument? *Studies in History and Philosophy of Science*, **22** (1): 23–36.

Cartwright, N. 1983. *How the Laws of Physics Lie*. Oxford: Oxford University Press.

Cartwright, N. and R. Frigg. 2007. String Theory under Scrutiny. *Physics World*, 20 (9): 14.

Chakravartty, A. 2007. *A Metaphysics for Scientific Realism: Knowing the Unobservable*. Cambridge: Cambridge University Press.

2008. What You Don't Know Can't Hurt You: Realism and the Unconceived. *Philosophical Studies*, **137** (1): 149–158.

Chalmers, A. 2011. Drawing Philosophical Lessons from Perrin's Experiments on Brownian Motion: A Response to van Fraassen. *British Journal for the Philosophy of Science*, 62 (4): 711–732.

Chang, H. 2004. *Inventing Temperature: Measurement and Scientific Progress*. Oxford: Oxford University Press.

Churchland, P. M. and C. A. Hooker, eds. 1985. *Images of Science: Essays on Realism and Empiricism*. Chicago, IL: University of Chicago Press.

Clark, P. 1976. Atomism versus Theromodynamics. In *Method and Appraisal in the Physical Sciences: The Critical Background to Modern Science, 1800–1905*, C. Howson (ed.), Cambridge: Cambridge University Press.

Cochran, W., F. Crick, and V. Vand. 1952. The Structure of Synthetic Polypeptides. I. The Transform of Atoms on a Helix. *Acta Crystallographica*, **5** (5): 581–586.

Collins, H. M. and T. J. Pinch. 1998. *The Golem: What You Should Know about Science*. Cambridge: Cambridge University Press.

Copernicus, N. 1543/1992. *On the Revolutions* (with a commentary by E. Rosen). Baltimore, MD: Johns Hopkins University Press.

Crick, F. 1954. The Structure of the Hereditary Material. *Scientific American*, 191 (October): 54–61.

1988. *What Mad Pursuit: A Personal View of Science*. New York, NY: Basic Books.

Curtis, W. E. 1914. Wave-Lengths of Hydrogen Lines and Determination of the Series Constant. *Proceedings of the Royal Society of London. Series A*, **90** (622): 605–620.

Cushing, J. T. 1981. Electromagnetic Mass, Relativity, and the Kaufmann Experiments. *American Journal of Physics*, **49**: 1133–1149.

Dawid, R. 2013. *String Theory and the Scientific Method*. Cambridge: Cambridge University Press.

de Regt, H. W. 1996. Philosophy and the Kinetic Theory of Gases. *British Journal for the Philosophy of Science*, **47** (1): 31–62.

2001. Spacetime Visualisation and the Intelligibility of Physical Theories. *Studies in History and Philosophy of Science Part B: Studies In History and Philosophy of Modern Physics*, **32** (2): 243–265.

2004. Discussion Note: Making Sense of Understanding. *Philosophy of Science*, **71** (1): 98–109.

de Regt, H. W. and D. Dieks. 2005. A Contextual Approach to Scientific Understanding. *Synthese*, **144** (1): 137–170.

De Rújula, A., H. Georgi, S. L. Glashow, and H. R. Quinn. 1974. Fact and Fancy in Neutrino Physics. *Reviews of Modern Physics*, **46**: 391–407.

Devitt, M. 2011. Are Unconceived Alternatives a Problem for Scientific Realism? *Journal for General Philosophy of Science*, **42** (2): 285–293.

Dicken, P. 2013. Normativity, the Base-Rate Fallacy, and Some Problems for Retail Realism. *Studies in History and Philosophy of Science Part A*, **44** (4): 563–570.

Donovan, A., L. Laudan, and R. Laudan. 1988. *Scrutinizing Science: Empirical Studies of Scientific Change*, Vol. 193. Baltimore, MD: John Hopkins University Press.

Douglas, H. 2009. *Science, Policy, and the Value-Free Ideal*. Pittsburgh, PA: University of Pittsburgh Press.

Dupré, J. 1993. *The Disorder of Things: Metaphysical Foundations of the Disunity of Science*. Cambridge, MA: Harvard University Press.

Dyson, F. W., A. S. Eddington, and C. Davidson. 1920. A Determination of the Deflection of Light by the Sun's Gravitational Field, from Observations Made at the Total Eclipse of May 29, 1919. *Philosophical Transactions of the Royal Society of London. Series A*, **220** (571–581): 291–333.

Earman, J. 2000. *Hume's Abject Failure: The Argument against Miracles*. Oxford: Oxford University Press.

Earman, J. and C. Glymour. 1980. Relativity and Eclipses: The British Eclipse Expeditions of 1919 and Their Predecessors. *Historical Studies in the Physical Sciences*, **11** (1): 49–85.

Eckert, M. and K. Märker. 2000. *Arnold Sommerfeld. Wissenschaftlicher Briefwechsel. Band 1: 1892–1918*. Munich: Deutsches Museum Verlag für Geschichte der Naturwissenschaften und der Technik.

Egg, M. 2016. Expanding Our Grasp: Causal Knowledge and the Problem of Unconceived Alternatives. *British Journal for the Philosophy of Science*, **67** (1): 115–141.

Einstein, A. 1907. Über das Relativistätsprinzip und die aus demselben gezogenen Folgerungen. *Jahrbuch der Radioaktivität und Elektronik*, **4**: 411–62.

1982. *Ideas and Opinions*. New York, NY: Crown Publishers Inc.

Einstein, A., H. Lorentz, H. Weyl, and H. Minkowski. 1952. *The Principle of Relativity*. New York, NY: Dover.

Elkin, L. O. 2003. Rosalind Franklin and the Double Helix. *Physics Today*, **56** (3): 42–48.

Ellis, G. and J. Silk. 2014. Scientific Method: Defend the Integrity of Physics. *Nature*, **516** (18 December 2014): 321-323.

Everitt, C. W. F. 1980. Experimental Tests of General Relativity: Past, Present and Future. In *Physics and Contemporary Needs*, Vol. 4, Riazuddin (ed.), Boston, MA: Springer US, 529–555.

Fahrbach, L. 2011. How the Growth of Science Ends Theory Change. *Synthese*, **180** (2): 139–155.

Feng, J. L. 2013. Naturalness and the Status of Supersymmetry. *Annual Review of Nuclear and Particle Science*, **63**: 351–382.

Feyerabend, P. 1975. *Against Method*. London: Verso.

Feynman, R. P. and M. Gell-Mann. 1958. Theory of the Fermi Interaction. *Physical Review*, **109** (1): 193.

Fine, A. 1986. Unnatural Attitudes: Realist and Instrumentalist Attachments to Science. *Mind*, **95** (378): 149–179.

 1991. Piecemeal Realism. *Philosophical Studies*, **61** (1): 79–96.

Fitzpatrick, S. 2013. Simplicity in the Philosophy of Science. Internet Encyclopedia of Philosophy, www.iep.utm.edu/simplici/.

Forman, P. 1968. The Doublet Riddle and Atomic Physics circa 1924. *Isis*: 156–174.

Forster, M. and E. Sober. 1994. How to Tell When Simpler, More Unified, or Less Ad Hoc Theories Will Provide More Accurate Predictions. *British Journal for the Philosophy of Science*, **45** (1): 1–35.

Frankel, H. 1982. The Development, Reception, and Acceptance of the Vine-Matthews-Morley Hypothesis. *Historical Studies in the Physical Sciences*, **13** (1): 1–39.

 2012. *The Continental Drift Controversy: Evolution into Plate Tectonics*. Cambridge: Cambridge University Press.

Franklin, R. E. and R. G. Gosling. 1953. The Structure of Sodium Thymonucleate Fibres. I. The Influence of Water Content. *Acta Crystallographica*, **6** (8–9): 673–677.

Friederich, S., R.V. Harlander, and K. Karaca. 2014. Philosophical Perspectives on Ad Hoc Hypotheses and the Higgs Mechanism. *Synthese*, **191** (16): 3897–3917.

Friedman, M. 1974. Explanation and Scientific Understanding. *Journal of Philosophy*, **71** (1): 5–19.

Frigg, R. and S. Hartmann. 2012. Models in Science. *Stanford Encyclopedia of Philosophy* (fall 2012 edition), E. N. Zalta (ed.), http://plato.stanford.edu/archives/fall2012/entries/models-science/.

Frigg, R. and I. Votsis. 2011. Everything You Always Wanted to Know about Structural Realism but Were Afraid to Ask. *European Journal for Philosophy of Science*, **1** (2): 227–276.

Frisch, M. 2005. *Inconsistency, Asymmetry, and Non-Locality: A Philosophical Investigation of Classical Electrodynamics*. Oxford: Oxford University Press.

2015. Predictivism and Old Evidence: A Critical Look at Climate Model Tuning. *European Journal for Philosophy of Science*, 5 (2): 171–190.

Galison, P. 1983. How the First Neutral-Current Experiments Ended. *Reviews of Modern Physics*, 55 (2): 477.

1987. *How Experiments End*. Chicago, IL: University of Chicago Press.

Gamow, G. 1952. *The Creation of the Universe*. 1st ed. London: MacMillan and Co.

1954. Modern Cosmology. *Scientific American*, 190 (3): 54–63.

1961. *The Creation of the Universe*. 2nd ed. (1st ed. 1952) London: MacMillan and Co.

Gardner, M.R. 1982. Predicting Novel Facts. *British Journal for the Philosophy of Science*, 33 (1): 1–15.

Georgi, H. and S. L. Glashow. 1972. Unified Weak and Electromagnetic Interactions without Neutral Currents. *Physical Review Letters*, 28 (22): 1494–1497.

1974. Unity of all Elementary-Particle Forces. *Physical Review Letters*, 32 (8): 438–441.

Giere, R. N. 1973. History and Philosophy of Science: Marriage of Convenience or Intimate Relationship. *British Journal for the Philosophy of Science*, 24 (3): 282–297.

1985. Philosophy of Science Naturalized. *Philosophy of Science*, 52: 331–356.

Gijsbers, V. 2007. Why Unification Is Neither Necessary Nor Sufficient for Explanation. *Philosophy of Science*, 74 (4): 481–500.

Gingerich, O. 1975. 'Crisis' versus Aesthetic in the Copernican Revolution. *Vistas in Astronomy*, 17 (1): 85–95.

Glashow, S. L. 1961. Partial-Symmetries of Weak Interactions. *Nuclear Physics*, 22: 579–588.

1980. Toward a Unified Theory: Threads in a Tapestry. *Science*, 210 (4476): 1319–1323.

Glymour, C. 1980. *Theory and Evidence*. Princeton, NJ: Princeton University Press.

Goldhaber, G., F. Pierre, G. Abrams, et al. 1976. Observation in e+ e− Annihilation of a Narrow State at 1865 MeV/c 2 Decaying to K π and K π π π. *Physical Review Letters*, 37 (5): 255–259.

Goodman, N. 1972. Seven Strictures on Similarity. In *Problems and Projects*, Indianapolis, IN: Bobs-Merril.

Gould, S. J. 1976. This View of Life: Darwin's Untimely Burial. *Natural History*, 85 (8): 24p.

Grant, R. 1852. *History of Physical Astronomy*. London: Henry G. Bohn.

Grosser, M. 1962. The Discovery of Neptune. Cambridge, MA: Harvard University Press.

Grünbaum, A. 1959. The Falsifiability of the Lorentz-Fitzgerald Contraction Hypothesis. *British Journal for the Philosophy of Science*, 10 (37): 48–50.

1976. Ad Hoc Auxiliary Hypotheses and Falsificationism. *British Journal for the Philosophy of Science*, **27** (4): 329–362.

1979. Is Freudian Psychoanalytic Theory Pseudo-Scientific by Karl Popper's Criterion of Demarcation? *American Philosophical Quarterly*, **16** (2): 131–141.

Hacking, I. 1982. Experimentation and Scientific Realism. *Philosophical Topics*, **13** (1): 71–87.

1983. *Representing and Intervening*. Cambridge: Cambridge Univiersity Press.

Hanson, N. R. 1962. The Irrelevance of History of Science to Philosophy of Science to Philosophy of Science. *Journal of Philosophy*, **59** (21): 574–586.

Harker, D. 2006. Accommodation and Prediction: The Case of the Persistent Head. *British Journal for the Philosophy of Science*, **57** (2): 309–321.

2008. On the Predilections for Predictions. *British Journal for the Philosophy of Science*, **59** (3): 429.

2013. How to Split a Theory: Defending Selective Realism and Convergence without Proximity. *British Journal for the Philosophy of Science*, **64** (1): 79–106.

Harre, R. 1960. An Introduction to the Logic of the Sciences. London: MacMillan.

Heilprin, J. 2013. Higgs Boson Discovery Confirmed after Physicists Review Large Hadron Collider Data At CERN. *Huffington Post*, 14/03/13. www .huffingtonpost.com/2013/03/14/higgs-boson-discovery-confirmed-cern-large-hadron-collider_n_2874975.html.

Held, C. 2011. Truth Does Not Explain Predictive Success. *Analysis*, **71** (2), 232–234.

Henderson, L. 2015. The No Miracles Argument and the Base Rate Fallacy. *Synthese*, **4** (194): 1295–1302.

Hess, H. H. 1962. History of Ocean Basins. *Petrologic Studies*, **4**: 599–620.

Hitchcock, C. and E. Sober. 2004. Prediction versus Accommodation and the Risk of Overfitting. *British Journal for the Philosophy of Science*, **55** (1): 1–34.

Holton, G. J. 1969. Einstein, Michelson, and the 'Crucial' Experiment. *Isis*, **60** (2): 133–197.

1973. *Thematic Origins of Scientific Thought: Kepler to Einstein*. Cambridge, MA: Harvard University Press.

Hon, G. 1995. Is the Identification of Experimental Error Contextually Dependent? The Case of Kaufmann's Experiment and Its Varied Reception. In *Scientific Practice: Theories and Stories of Doing Physics*, J. Buchwald (ed.), Chicago, IL: University of Chicago Press, 170–223.

Howson, C. 2000. *Hume's Problem: Induction and the Justification of Belief*. Oxford: Clarendon Press.

2013. Exhuming the No-Miracles Argument. *Analysis*, **73** (2): 205–211.

Howson, C. and A. Franklin. 1991. Maher, Mendeleev and Bayesianism. *Philosophy of Science*, **58** (4): 574–585.

Howson, C. and P. Urbach. 2006. *Scientific Reasoning: The Bayesian Approach*. LaSalle, IL: Open Court Publishing.

Hoyer, U. 1981. *Work on Atomic Physics (1912–1917)*, Vol. 2, Niels Bohr Collected Works. Amsterdam: North-Holland.

Hoyle, F. 1948. A New Model for the Expanding Universe. *Monthly Notices of the Royal Astronomical Society*, **108**: 372.

1949. March 11 Meeting of the Royal Astronomical Society. *The Observatory*, **69**: 47–54.

1955. *Frontiers of Astronomy*. London: Heinemann Education Books Limited.

Hoyningen-Huene, P. 2013. *Systematicity: The Nature of Science*. Oxford: Oxford University Press.

Hudson, R. G. 2007. What's Really at Issue with Novel Predictions? *Synthese*, **155** (1): 1–20.

2013. *Seeing Things: The Philosophy of Reliable Observation*. Oxford: Oxford University Press.

Hunt, J. C. 2012. On Ad Hoc Hypotheses. *Philosophy of Science*, **79** (1): 1–14.

Irzik, G. and R. Nola. 2011. A Family Resemblance Approach to the Nature of Science for Science Education. *Science & Education*, **20** (7–8): 591–607.

Ivanova, M. 2010. Pierre Duhem's Good Sense as a Guide to Theory Choice. *Studies in History and Philosophy of Science Part A*, **41** (1): 58–64.

Jammer, M. 1989. *The Conceptual Development of Quantum Mechanics*. Los Angeles, CA: Tomash Publishers.

Janssen, M. 2002a. COI Stories: Explanation and Evidence in the History of Science. *Perspectives on Science*, **10** (4): 457–522.

2002b. Reconsidering a Scientific Revolution: The Case of Einstein versus Lorentz. *Physics in Perspective*, **4** (4): 421–446.

Jansson, L. and J. Tallant. 2017. Quantitative Parsimony: Probably for the Better. *British Journal for the Philosophy of Science*, **68** (3): 781–803.

Jeffreys, H. 1973. *Scientific Inference*. Cambridge: Cambridge University Press.

Johansson, L. G. and K. Matsubara. 2011. String Theory and General Methodology: A Mutual Evaluation. *Studies In History and Philosophy of Science Part B: Studies in History and Philosophy of Modern Physics*, **42** (3): 199–210.

Judson, H. F. 1996. *The Eighth Day of Creation: Makers of the Revolution in Biology*. New York, NY: Cold Spring Harbor Laboratory Press.

Kelly, T. 2003. Epistemic Rationality as Instrumental Rationality: A Critique. *Philosophy and Phenomenological Research*, **66** (3): 612–640.

Kemeny, J. G. and P. Oppenheim. 1952. Degree of Factual Support. *Philosophy of Science*, **19** (4): 307–324.

Kennefick, D. 2007. Not Only because of Theory: Dyson, Eddington and the Competing Myths of the 1919 Eclipse Expedition. *Arxiv preprint:* arXiv:0709.0685.

Khalifa, K. 2012. Inaugurating Understanding or Repackaging Explanation? *Philosophy of Science*, **79** (1): 15–37.

2013. The Role of Explanation in Understanding. *British Journal for the Philosophy of Science*, **64** (1): 161–187.

Kinzel, K. 2015. Narrative and Evidence. How Can Case Studies from the History of Science Support Claims in the Philosophy of Science? *Studies in History and Philosophy of Science Part A*, **49**: 48–57.

Kitcher, P. 1976. Explanation, Conjunction, and Unification. *Journal of Philosophy*, **73** (8): 207–212.

  1981. Explanatory Unification. *Philosophy of Science*, **48** (4): 507–531.

  1993. *The Advancement of Science*. Oxford: Oxford University Press.

Klug, A. 1968. Rosalind Franklin and the Discovery of the Structure of DNA. *Nature*, **219** (24 August): 808–844.

  2004. The Discovery of the DNA Double Helix. *Journal of Molecular Biology*, **335** (1): 3–26.

Koester, D., D. Sullivan, and D. H. White. 1982. Theory Selection in Particle Physics: A Quantitative Case Study of the Evolution of Weak-Electromagnetic Unification Theory. *Social Studies of Science*, **12** (1): 73–100.

Kragh, H. 1985. The Fine Structure of Hydrogen and the Gross Structure of the Physics Community, 1916–26. *Historical Studies in the Physical Sciences*, **15** (2): 67–125.

  1996. *Cosmology and Controversy: The Historical Development of Two Theories of the Universe*. Princeton, NJ: Princeton University Press.

  1999. Steady-State Cosmology and General Relativity: Reconciliation or Conflict? In *The Expanding Worlds of General Relativity*, H. Goenner, J. Renn, J. Ritter, and T. Sauer (eds.), Boston, MA: Birkhäuser, 377–404.

  2012. *Niels Bohr and the Quantum Atom: The Bohr Model of Atomic Structure 1913–1925*. Oxford: Oxford University Press.

Krämer, M. 2013. The Landscape of New Physics. *The Guardian*, January 9, 2013. www.theguardian.com/science/life-and-physics/2013/jan/09/physics-particlephysics.

Kuhn, T. S. 1957. *The Copernican Revolution: Planetary Astronomy in the Development of Western Thought*. Cambridge, MA: Harvard University Press.

  1962/1996. *The Structure of Scientific Revolutions*. Chicago, IL: University of Chicago Press.

  1970a. Logic of Discovery or Psychology of Research. In *Criticism and the Growth of Knowledge, Proceedings of the International Colloquium in the Philosophy of Science*, I. Lakatos and A. Musgrave (eds.), Cambridge: Cambridge University Press, 1–24.

  1970b. Notes on Lakatos. *PSA: Proceedings of the Biennial Meeting of the Philosophy of Science Association*, 137–146.

  1970c. Reflections on My Critics. In *Criticism and the Growth of Knowledge, Proceedings of the International Colloquium in the Philosophy of Science*, I. Lakatos and A. Musgrave (eds.), Cambridge: Cambridge University Press, 231–278.

  1977a. Objectivity, Value Judgment, and Theory Choice. In *The Essential Tension*, Chicago, IL: University of Chicago Press, 320–333.

1977b. The Relations between the History and the Philosophy of Science. In *The Essential Tension*. Chicago, IL: University of Chicago Press, 3–20.

1987. *Black-Body Theory and the Quantum Discontinuity, 1894–1912*. Chicago, IL: University of Chicago Press.

1990. Dubbing and Redubbing: The Vulnerability of Rigid Designation. *Minnesota Studies in the Philosophy of Science*, **14**: 298–318.

Kukla, A. 1996. Does Every Theory Have Empirically Equivalent Rivals? *Erkenntnis*, **44**: 137–166.

Ladyman, J. 1998. What Is Structural Realism? *Studies in History and Philosophy of Science Part A*, **29** (3): 409–424.

1999. Review: A Novel Defense of Scientific Realism. Jarrett Leplin. *British Journal for the Philosophy of Science*, **50** (1): 181–188.

2011. Structural Realism versus Standard Scientific Realism: The Case of Phlogiston and Dephlogisticated Air. *Synthese*, **180**: 87–101.

Ladyman, J., I. Douven, L. Horsten, and B. Fraassen. 1997. A Defence of van Fraassen's Critique of Abductive Inference: Reply to Psillos. *Philosophical Quarterly*, **47** (188): 305–321.

Lakatos, I. 1970a. Falsification and the Methodology of Scientific Research Programmes. In *Criticism and the Growth of Knowledge*, I. Lakatos and A. Musgrave (eds.), Cambridge: Cambridge University Press, 91–196.

1970b. History of Science and Its Rational Reconstructions. PSA: Proceedings of the Biennial Meeting of the Philosophy of Science Association, 1970, 91–136.

1976. *Proofs and Refutations*. Cambridge: Cambridge University Press.

1978. *The Methodology of Scientific Research Programmes: Volume 1: Philosophical Papers*. J. Worrall and G. Currie (eds.). Cambridge: Cambridge University Press.

Lange, M. 2001. The Apparent Superiority of Prediction to Accommodation as a Side Effect: A Reply to Maher. *British Journal for the Philosophy of Science*, **52** (3): 575–588.

2002. Baseball, Pessimistic Inductions and the Turnover Fallacy. *Analysis*, **62** (276): 281–285.

LaPorte, J. 2009. *Natural Kinds and Conceptual Change*. Cambridge: Cambridge University Press.

Laudan, L. 1977. *Progress and Its Problems: Towards a Theory of Scientific Growth*. Berkeley: University of California Press.

1981. A Confutation of Convergent Realism. *Philosophy of Science*, **48** (1): 19–49.

1982. Commentary: Science at the Bar – Causes for Concern. *Science, Technology & Human Values*, 7 (4): 16–19.

1983. The Demise of the Demarcation Problem. In *Physics, Philosophy and Psychoanalysis: Essays in Honour of A. Grünbaum*, R. S. Cohen and L. Laudan (eds.), Dordrecht: Reidel, 111–127.

1986. Some Problems Facing Intuitionist Meta-Methodologies. *Synthese*, **67** (1): 115–129.

1987a. Progress or Rationality? The Prospects for Normative Naturalism. *American Philosophical Quarterly*, **24** (1): 19–31.

1987b. Relativism, Naturalism and Reticulation. *Synthese*, **71** (3): 221–234.

1989. If It Ain't Broke, Don't Fix It. *British Journal for the Philosophy of Science*, **40** (3): 369–375.

1990. Normative Naturalism. *Philosophy of Science*, **57** (1): 44–59.

Laudan, L., A. Donovan, R. Laudan, et al. 1986. Scientific Change: Philosophical Models and Historical Research. *Synthese*, **69** (2): 141–223.

Laudan, L. and J. Leplin. 1991. Empirical Equivalence and Underdetermination. *Journal of Philosophy*, **88** (9): 449–472.

Lee, W.-Y. 2013. Akaike's Theorem and Weak Predictivism in Science. *Studies in History and Philosophy of Science Part A*, **44** (4): 594–599.

Leite, A. 2007. Epistemic Instrumentalism and Reasons for Belief: A Reply to Tom Kelly's 'Epistemic Rationality as Instrumental Rationality: A Critique'. *Philosophy and Phenomenological Research*, **75** (2): 456–464.

Leplin, J. 1975. The Concept of an Ad Hoc Hypothesis. *Studies in History and Philosophy of Science Part A*, **5** (4): 309–345.

1982. The Assessment of Auxiliary Hypotheses. *British Journal for the Philosophy of Science*, **33** (3): 235–249.

1984. *Scientific Realism*. Berkeley: University of California Press.

1997. A Novel Defense of Scientific Realism. Oxford: Oxford University Press.

Lewis, C. I. 1946. *An Analysis of Knowledge and Valuation*. LaSalle, IL: Open Court.

Lewis, D. 1973. *Counterfactuals*. Oxford: Blackwell.

1986. *Philosophical Papers*. Oxford: Oxford University Press.

Lewis, P. J. 2001. Why the Pessimistic Induction Is a Fallacy. *Synthese*, **129** (3): 371–380.

Lipton, P. 1991/2004. *Inference to the Best Explanation*. London: Routledge.

Lycan, W. G. 2002. Explanation and Epistemology. In *The Oxford Handbook of Epistemology*, P. K. Moser (ed.), Oxford: Oxford University Press, 408–433.

Lyons, T. D. 2006. Scientific Realism and the Stratagema de Divide et Impera. *British Journal for the Philosophy of Science*, **57** (3): 537–560.

Maddox, B. 2002. *Rosalind Franklin: The Dark Lady of DNA*. New York, NY: HarperCollins.

Magnus, P. 2010. Inductions, Red Herrings, and the Best Explanation for the Mixed Record of Science. *British Journal for the Philosophy of Science*, **61** (4): 803–819.

Magnus, P. and C. Callender. 2004. Realist Ennui and the Base Rate Fallacy. *Philosophy of Science*, **71** (3): 320–338.

Maher, P. 1988. Prediction, Accommodation, and the Logic of Discovery. *PSA: Proceedings of the 1988 Biennial Meeting of the Philosophy of Science Association*, A. Fine and J. Leplin (eds.), Vol. 1: *Contributed Papers*, 273–285.

Mathews, C. K., K. E. van Holde, and K. G. Ahern. 2000. *Biochemistry*. 3rd ed. New York, NY: Pearson Education.

Maxwell, N. 2002. Karl Raimund Popper. In *British Philosophers, 1800–2000*, L. McHenry, P. Dematteis, and P. Fosl (eds.), Columbia, SC: Bruccoli Clark Layman, 176–194.

Mayo, D. G. 1996. *Error and the Growth of Experimental Knowledge*. Chicago, IL: University of Chicago Press.

McAllister, J. W. 1997. Phenomena and Patterns in Data Sets. *Erkenntnis*, **47**: 217–228.

1999. *Beauty and Revolution in Science*: Ithaca, NY: Cornell University Press.

McIntyre, L. 2001. Accommodation, Prediction, and Confirmation. *Perspectives on Science*, **9** (3): 308–323.

McMullin, E. 1968. What Do Physical Models Tell Us? *Studies in Logic and the Foundations of Mathematics*, **52**: 385–396.

1974. History and Philosophy of Science: A Marriage of Convenience? In R. S. Cohen et al. (eds.) *PSA 1974: Boston Studies in the Philosophy of Science*, Vol. 32. Dordrecht: Springer, 585–601.

1976. The Fertility of Theory and the Unit for Appraisal in Science. In *Essays in the Memory of Imre Lakatos*, R. S. Cohen (ed.), Dordrecht: D. Reidel Publishing Company, 395–432.

1983. Values in Science. PSA: Proceedings of the Biennial Meeting of the Philosophy of Science Association, P. Asquith and T. Nickles (eds.), Vol. 2: *Symposia and Invited Papers*, 3–28.

1984. A Case for Scientific Realism. In *Scientific Realism*, J. Leplin (ed.), Berkeley: University of California Press, 8–41.

1985. Galilean Idealization. *Studies in History and Philosophy of Science Part A*, **16** (3): 247–273.

1998. Rationality and Paradigm Change in Science. In *Philosophy of Science: The Central Issues*, M. Curd and J. Cover (eds.), New York, NY: W. W. Norton & Company, 55–78.

McVittie, G. C. 1949. March 11 Meeting of the Royal Astronomical Society. *The Observatory*, **69**: 47–54.

1951. The Cosmological Problem. *Science News*, **21**: 61–75.

Mehra, J. and H. Rechenberg. 1982. *The Historical Development of Quantum Theory*. Vol. 1 (2 vols.). New York, NY: Springer.

Mendeleev, D. 1879. The Periodic Law of the Chemical Elements. *Chemical News*, **40**.

1889. The Periodic Law of the Chemical Elements. *Journal of the Chemical Society*, **55**: 634–56.

1901. *The Principles of Chemistry*. (6th ed.; trans. from Russian by George Kamensky; A. J. Greenaway, ed.). New York, NY: Collier.

Menke, C. 2014. Does the Miracle Argument Embody a Base Rate Fallacy? *Studies in History and Philosophy of Science Part A*, **45** (March), 103–108.

Meyer, L. 1870. Die Natur der chemischen Elemente als Funktion ihrer Atomgewichte. *Supplements to Justig Liebig's Annalen der Chemie*, **VII**: 354.

Mill, J. S. 1867. *An Examination of Sir William Hamilton's Philosophy: And of the Principal Philosophical Questions Discussed in His Writings.* London: Walter Scott.

Miller, A. I. 1981. *Albert Einstein's Special Theory of Relativity: Emergence (1905) and Early Interpretation (1905–1911).* Reading, MA: Addison-Wesley.

Miller, D. 2014. *Representing Space in the Scientific Revolution.* Cambridge: Cambridge University Press.

Milne, E. A. 1949. March 11 Meeting of the Royal Astronomical Society. *The Observatory,* **69**: 47–54.

Morganti, M. 2011. Truth and Success: Reply to Held. *The Reasoner,* **5** (7): 106–7.

2012. Truth and Success, Again: Reply to Held on Generalist versus Particularist (Anti-)Realism. *The Reasoner,* **6** (6): 99–100.

Morrison, M. 2000. *Unifying Scientific Theories.* Cambridge: Cambridge University Press.

Musgrave, A. 1974. Logical versus Historical Theories of Confirmation. *British Journal for the Philosophy of Science,* 25 (1): 1–23.

1988. The Ultimate Argument for Scientific Realism. In *Relativism and Realism in Science,* R. Nola (ed.), Dordrecht: Kluwer Academic Publishers, 229–252.

Nolan, D. 1997. Quantitative Parsimony. *British Journal for the Philosophy of Science,* **48** (3): 329–343.

1999. Is Fertility Virtuous in Its Own Right? *British Journal for the Philosophy of Science,* **50** (2): 265–282.

Norton, J. D. 1987. The Logical Inconsistency of the Old Quantum Theory of Black Body Radiation. *Philosophy of Science,* **54** (3): 327–350.

2002. A Paradox in Newtonian Gravitation Theory II. In *Inconsistency in Science,* J. Meheus (ed.), Dordrecht: Kluwer Academic Publishers, 185–195.

Nyhof, J. 1988. Philosophical Objections to the Kinetic Theory. *British Journal for the Philosophy of Science,* **39** (1): 81–109.

Okasha, S. 2011. Theory Choice and Social Choice: Kuhn versus Arrow. *Mind,* **120** (477): 83–115.

Olby, R. C. 1974. *The Path to the Double Helix: The Discovery of DNA.* New York, NY: Courier Corporation.

Olsson, E. 2012. Coherentist Theories of Epistemic Justification. In *Stanford Encyclopedia of Philosophy* (spring 2013 edition), E. N. Zalta (ed.), http://plato.stanford.edu/archives/spr2013/entries/justep-coherence/.

Oreskes, N. 2003. *Plate Tectonics: An Insider's History of the Modern Theory of the Earth.* Boulder, CO: Westview Press.

Overbye, D. 2011. Scientists Report Second Sighting of Faster-than-Light Neutrinos. *New York Times,* 18/11/2011. www.nytimes.com/2011/11/19/science/space/neutrino-finding-is-confirmed-in-second-experiment-opera-scientists-say.html?partner=rss&emc=rss.

2012. 'Physicists Find Elusive Particle Seen as Key to Universe.' *New York Times,* 05/07/2012. www.nytimes.com/2012/07/05/science/cern-physicists-may-have-discovered-higgs-boson-particle.html?_r=2&pagewanted=all&.

Pais, A. 1991. *Neils Bohr's Times: In Physics, Philosophy, and Polity.* Oxford: Clarendon Press.

Palter, R. 1970. An Approach to the History of Early Astronomy. *Studies in History and Philosophy of Science Part A*, **1** (2): 93–133.

Papineau, D. 2015. Naturalism. In *Stanford Encyclopedia of Philosophy* (fall 2015 edition), E. N. Zalta (ed.), http://plato.stanford.edu/archives/spr2009/entries/naturalism/.

Paschen, F. 1916. Bohr's Heliumlinien. *Annalen der Physik*, **355** (16): 901–940.

Pashby, T. 2012. Dirac's Prediction of the Positron: A Case Study for the Current Realism Debate. *Perspectives on Science*, **20** (4): 440–475.

Pauli, W. 1946. Lecture: Exclusion Principle and Quantum Mechanics. Nobelprize.org. Nobel Media AB 2014, accessed 13 Feb. www.nobelprize.org/nobel_prizes/physics/laureates/1945/pauli-lecture.html.

　1979. *Wissenschaftlicher Briefwechsel mit Bohr, Einstein, Heisenberg, ua, Bd. 1, 1919–1924.* New York, NY: Springer.

Peters, D. 2014. What Elements of Successful Scientific Theories Are the Correct Targets for 'Selective' Scientific Realism? *Philosophy of Science*, **81** (3): 377–397.

Pickering, A. 1984a. Against Putting the Phenomena First: The Discovery of the Weak Neutral Current. *Studies in History and Philosophy of Science*, **15** (2): 85–117.

　1984b. *Constructing Quarks: A Sociological History of Particle Physics.* Chicago, IL: University of Chicago Press.

Pigliucci, M. 2013. The Demarcation Problem: A (Belated) Response to Laudan. In *Philosophy of Pseudoscience: Reconsidering the Demarcation Problem*, M. Pigliucci and M. Boudry (eds.), Chicago, IL: University of Chicago Press.

Pigliucci, M. and M. Boudry. 2013. *Philosophy of Pseudoscience: Reconsidering the Demarcation Problem.* Chicago, IL: University of Chicago Press.

Poincaré, H. 1914. *Science and Method*, trans. by F. Maitland, London: Thomas Nelson.

Pompa, L. 1967. Family Resemblance. *Philosophical Quarterly*, **17** (66): 63–69.

Popper, K. R. 1940. What Is Dialectic? *Mind*, **49** (196): 403–426.

　1959a. *The Logic of Scientific Discovery.* London: Routledge.

　1959b. Testability and 'Ad-Hocness' of the Contraction Hypothesis. *British Journal for the Philosophy of Science*, **10** (37): 50.

　1963/1978. *Conjectures and Refutations: The Growth of Scientific Knowledge.* 4th ed. London: Butler & Tanner Limited.

　1966. A Note on the Difference between the Lorentz-Fitzgerald Contraction and the Einstein Contraction. *British Journal for the Philosophy of Science*, **16** (64): 332–333.

　1970. Normal Science and Its Dangers. In *Criticism and the Growth of Knowledge, Proceedings of the International Colloquium in the Philosophy of Science*, I. Lakatos and A. Musgrave (eds.), Cambridge: Cambridge University Press, 51–58.

　1976. *Unended Quest: An Intellectual Autobiography.* London: Routledge.

1978. Natural Selection and the Emergence of Mind. *Dialectica*, **32** (3–4): 339–355.

Popper, K. R. and P. A. Schilpp. 1974. *The Philosophy of Karl Popper*. LaSalle, IL: Open Court.

Psillos, S. 1999. *Scientific Realism: How Science Tracks Truth*. London: Routledge.

2001. Predictive Similarity and the Success of Science: A Reply to Stanford. *Philosophy of Science*, **68** (3): 346–355.

2009. *Knowing the Structure of Nature: Essays on Realism and Explanation*. London: Palgrave Macmillan.

2011. Making Contact with Molecules: On Perrin and Achinstein. In *Philosophy of Science Matters: The Philosophy of Peter Achinstein*, G. J. Morgan (ed.), Oxford: Oxford University Press, 177–190.

Putnam, H. 1979. *Philosophical Papers: Volume 1, Mathematics, Matter and Method*. Cambridge: Cambridge University Press.

Quine, W. V. 1969. *Ontological Relativity and Other Essays*. New York, NY: Columbia University Press.

Quine, W. V. and J. S. Ullian. 1970. *Web of Belief*. New York, NY: Random House.

Reich, E. S. 2011. Speedy Neutrinos Challenge Physicists, *Nature*, 477 (September 27): 520.

2012. Timing Glitches Dog Neutrino Claim: Team Admits to Possible Errors in Faster-than-Light Finding. *Nature*, **483** (7387): 17–18.

Richman, R. J. 1962. Something Common. *Journal of Philosophy*, **59** (26): 821–830.

Robotti, N. 1983. The Spectrum of ζ Puppis and the Historical Evolution of Empirical Data. *Historical Studies in the Physical Sciences*, **14** (1), 123–145.

Roush, S. 2010. Optimism about the Pessimistic Induction. In *New Waves in Philosophy of Science*, P. D. Magnus and J. Busch (eds.), London: Palgrave Macmillan.

Ruhmkorff, S. 2011. Some Difficulties for the Problem of Unconceived Alternatives. *Philosophy of Science*, **78** (5): 875–886.

Ruse, M. 1977. Karl Popper's Philosophy of Biology. *Philosophy of Science*, **44** (4): 638–661.

Saatsi, J. 2005a. On the Pessimistic Induction and Two Fallacies. *Philosophy of Science*, **72** (5): 1088–1098.

2005b. Reconsidering the Fresnel–Maxwell Theory Shift: How the Realist Can Have Her Cake and EAT It Too. *Studies in History and Philosophy of Science Part A*, **36** (3): 509–538.

2009. Form vs. Content-Driven Arguments for Realism. In *New Waves in Philosophy of Science*, P. D. Magnus and J. Busch (eds.), London: Palgrave Macmillan, 8–28.

in press. Historical Inductions, Old and New. *Synthese*, https://link.springer .com/article/10.1007/s11229-015-0855-5.

Saatsi, J. and P. Vickers. 2011. Miraculous Success? Inconsistency and Untruth in Kirchhoff's Diffraction Theory. *British Journal for the Philosophy of Science*, **62** (1): 29–46.

Sakurai, J. 1974. Remarks on Neutral Current Interactions. In *Neutrinos-1974, American Institute for Physics Conference Proceedings, 26–28 April 1974, Philadelphia*, C. Baltay (ed.), New York, NY: American Institute for Physics, 57–63.

    1978. Neutral Currents and Gauge Theories – Past, Present, and Future. In *Current Trends in the Theory of Fields (Tallahassee-1978): A Symposium in Honor of P. A. M. Dirac*, D. Lannutti and E. Williams (eds.), College Park, MD: American Institute of Physics, 38.

Salmon, W. 1984. *Scientific Explanation and Causal Structure of the World.* Princeton, NJ: Princeton University Press.

    1990. The Appraisal of Theories: Kuhn Meets Bayes. PSA: Proceedings of the Biennial Meeting of the Philosophy of Science Association, Vol. 2: *Symposia and Invited Papers*, 325–332.

Scerri, E. R. 2007. *The Periodic Table: Its Story and Its Significance.* Oxford: Oxford University Press.

Scerri, E. R. and J. Worrall. 2001. Prediction and the Periodic Table. *Studies in History and Philosophy of Science Part A*, **32** (3): 407–452.

Schaffner, K. F. 1974. Einstein versus Lorentz: Research Programmes and the Logic of Comparative Theory Evaluation. *British Journal for the Philosophy of Science*, **25** (1): 45–78.

Schindler, S. 2007. Rehabilitating Theory: Refusal of the 'Bottom-Up' Construction of Scientific Phenomena. *Studies in History and Philosophy of Science Part A*, **38** (1): 160–184.

    2008a. Model, Theory, and Evidence in the Discovery of the DNA Structure. *British Journal for the Philosophy of Science*, **59** (4): 619–658.

    2008b. Use-Novel Predictions and Mendeleev's Periodic Table: Response to Scerri and Worrall (2001). *Studies in History and Philosophy of Science Part A*, **39** (2): 265–269.

    2011. Bogen and Woodward's Data-Phenomena Distinction, Forms of Theory-Ladenness, and the Reliability of Data. *Synthese*, **182** (1): 39–55.

    2013a. The Kuhnian Mode of HPS. *Synthese*, **190** (18): 4137–4154.

    2013b. Theory-Laden Experimentation. *Studies in History and Philosophy of Science Part A*, **44** (1): 89–101.

    2014a. Novelty, Coherence, and Mendeleev's Periodic Table. *Studies in History and Philosophy of Science Part A*, **45** (March): 62–69.

    2014b. A Matter of Kuhnian Theory Choice. The GWS Model and the Neutral Current. *Perspectives on Science*, **22** (4): 491–522.

    2015. Scientific Discovery: That-Whats and What-Thats. *Ergo*, **2** (6): 123–148.

Schulmann, R., A. J. Kox, M. Janssen, and J. Illy, eds. 1998. *The Collected Papers of Albert Einstein. Vol. 8: The Berlin Years: Correspondence 1914–1918.* Princeton, NJ: Princeton University Press.

Schupbach, J.N. 2017. Experimental Explication. *Philosophy and Phenomenological Research*, **94** (3): 672–710.

Schurz, G. 2011. Structural Correspondence, Indirect Reference, and Partial Truth: Phlogiston Theory and Newtonian Mechanics. *Synthese*, 180 (2): 103–120.

Sciama, D. 1955. Evolutionary Processes in Cosmology. *Advancement of Science*, 12 (45): 38–42.

1959. *The Unity of the Universe*. London: Faber and Faber.

1960. Observational Aspects of Cosmology. *Vistas in Astronomy*, 3: 311–328.

1961. New Developments in Cosmology. *La Nuova Critica*, 11: 3–16.

Segall, R. 2008. Fertility and Scientific Realism. *British Journal for the Philosophy of Science*, 59 (2): 237–246.

Serwer, D. 1977. Unmechanischer Zwang: Pauli, Heisenberg, and the Rejection of the Mechanical Atom, 1923–1925. *Historical Studies in the Physical Sciences*, 8: 189–256.

Simon, M. A. 1969. When Is a Resemblance a Family Resemblance? *Mind*, 78 (311): 408–416.

Smith, J. R. 1976. Persistence and Periodicity: A Study of Mendeleev's Contribution to the Foundations of Chemistry. PhD dissertation, University of London.

Smolin, L. 2007. *The Trouble with Physics: The Rise of String Theory, the Fall of a Science, and What Comes Next*. Boston, MA: Houghton Mifflin Harcourt.

Sober, E. 1981. The Principle of Parsimony. *British Journal for the Philosophy of Science*, 32 (1), 145–156.

1990. Let's Razor Ockham's Razor. *Explanation and Its Limits, Royal Institute of Philosophy Supplementary*, 27: 73–94.

2001. Simplicity. In *A Companion to the Philosophy of Science*, W. H. Newton-Smith (ed.), Oxford: Blackwell, 433–442.

2008. *Evidence and Evolution: The Logic behind the Science*. Cambridge: Cambridge University Press.

2015. *Ockham's Razors*. Cambridge: Cambridge University Press.

Sommerfeld, A. 1916a. Zur quantentheorie der spektrallinien. *Annalen der Physik*, 356 (17): 1–94.

1916b. Zur Theorie der Balmerschen Serie. *Sitzungsberichte der Königlich Bayerischen Akademie der Wissenschaften Matematisch-physikalische Klasse*: 425–500.

1923. *Atomic Structure and Spectral Lines*. Trans. by Henry Brose. London: Methuen & Co. Ltd.

Stanford, P. K. 2000. An Antirealist Explanation of the Success of Science. *Philosophy of Science*, 67 (2): 266–284.

2001. Refusing the Devil's Bargain: What Kind of Underdetermination Should We Take Seriously? *Philosophy of Science*, 68 (3), Supplement: Proceedings of the 2000 Biennial Meeting of the Philosophy of Science Association. Part I: Contributed Papers: S1–S12.

2006. *Exceeding Our Grasp: Science, History, and the Problem of Unconceived Alternatives*. Oxford: Oxford University Press.

Steele, K. and C. Werndl. 2013. Climate Models, Calibration, and Confirmation. *British Journal for the Philosophy of Science*, **64** (3): 609–635.

2016. Model-Selection Theory: The Need for a More Nuanced Picture of Use-Novelty and Double-Counting. *British Journal for the Philosophy of Science*: axw024.

Stump, D. J. 2007. Pierre Duhem's Virtue Epistemology. *Studies in History and Philosophy of Science*, **18** (1): 149–159.

Swinburne, R. 1997. *Simplicity as Evidence of Truth*, Vol. 61, Aquinus Lecture. Milwaukee, WI: Marquette University Press.

2001. *Epistemic Justification*, Vol. 81. Oxford: Oxford University Press.

't Hooft, G. 1971. Renormalizable Lagrangians for Massive Yang-Mills Fields. *Nuclear Physics B*, **35** (1): 167–188.

1980. Gauge Theories of the Forces between Elementary Particles. *Scientific American*, **242**.

Taylor, C. 1967. The Patterson Function. *Physics Education*, **2** (5): 276.

Thomson, J. 1919. Joint Eclipse Meeting of the Royal Society and the Royal Astronomical Society, 6 November 1919. *The Observatory, London*, **42** (545): 389–398.

Thorburn, W. M. 1918. The Myth of Occam's Razor. *Mind*, **27** (107): 345–353.

Trout, J. D. 2002. Scientific Explanation and the Sense of Understanding. *Philosophy of Science*, **69** (2): 212–233.

van Fraassen, B. 1980. *The Scientific Image*. Oxford: Oxford University Press.

1989. *Laws and Symmetry*. Oxford: Oxford University Press.

2000. The False Hopes of Traditional Epistemology. *Philosophy and Phenomenological Research*, **60** (2): 253–280.

2009. The Perils of Perrin, in the Hands of Philosophers. *Philosophical Studies*, **143** (1): 5–24.

van Spronsen, J. W. 1969. The Priority Conflict between Mendeleev and Meyer. *Journal of Chemical Education*, **46** (3): 136.

Vickers, P. 2012. Historical Magic in Old Quantum Theory? *European Journal for Philosophy of Science*, **2** (1): 1–19.

2013a. A Confrontation of Convergent Realism. *Philosophy of Science*, **80** (2): 189–211.

2013b. *Understanding Inconsistent Science*. Oxford: Oxford University Press.

Vine, F. and D. Matthews. 1963. Magnetic Anomalies over Oceanic Ridges. *Nature*, **199** (4897): 947–949.

von Klüber, H. 1960. The Determination of Einstein's Light-Deflection in the Gravitational Field of the Sun. *Vistas in Astronomy*, **3**: 47–77.

Votsis, I. 2011. The Prospective Stance in Realism. *Philosophy of Science*, **78** (5): 1223–1234.

Warburg, E. 1915. *Die Physik: Die Kultur der Gegenwart*. Leipzig: Teubner.

Watson, J. 1968. *The Double Helix: A Personal Account of the Discovery of the Structure of DNA*. New York, IL: New American.

Weinberg, S. 1967. A Model of Leptons. *Physical Review Letters*, **19** (21): 1264.

1974. Recent Progress in Gauge Theories of the Weak, Electromagnetic and Strong Interactions. *Reviews of Modern Physics*, **46**: 255–277.

1979. Conceptual Foundations of the Unified Theory of Weak and Electromagnetic Interactions/Nobel Lecture, 8 December 1979.

1993. *Dreams of a Final Theory*. London: Vintage.

Wetterich, C. 2012. Where to Look for Solving the Gauge Hierarchy Problem? *Physics Letters B*, **718** (2): 573–576.

White, R. 2003. The Epistemic Advantage of Prediction over Accommodation. *Mind*, **112** (448): 653–683.

Wilkins, M. 2003. *The Third Man of the Double Helix: Memoirs of a Life in Science*. Oxford: Oxford University Press.

Will, C. M. 1993. *Was Einstein Right?: Putting General Relativity to the Test*, New York, NY: Basic Books.

Wittgenstein, L. 1953. *Philosophical Investigations*. G.E.M. Anscombe and R. Rhees (eds.), G.E.M. Anscombe (trans.), Oxford: Blackwell.

Woit, P. 2011. *Not Even Wrong: The Failure of String Theory and the Continuing Challenge to Unify the Laws of Physics*. London: Random House.

Woodward, J. 2003. *Making Things Happen: A Theory of Causal Explanation*. Oxford: Oxford University Press.

2014a. Scientific Explanation. In *Stanford Encyclopedia of Philosophy*, E. N. Zalta (ed.), http://plato.stanford.edu/archives/win2014/entries/scientific-explanation/.

2014b. Simplicity in the Best Systems Account of Laws of Nature. *British Journal for the Philosophy of Science*, **65** (1): 91–123.

Worrall, J. 1976. Thomas Young and the 'Refutation' of Newtonian Opticas: A Case-Study in the Interaction of Philosophy of Science and History of Science. In *Method and Appraisal in the Physical Sciences*, C. Howson (ed.), London: Cambridge University Press, 107–180.

1985. Scientific Discovery and Theory-Confirmation. In *Change and Progress in Modern Science*, J. C. Pitt (ed.), Dordrecht: D. Reidel, 301–331.

1988. The Value of a Fixed Methodology. *British Journal for the Philosophy of Science*, **39** (2): 263–275.

1989a. Fix It and Be Damned: A Reply to Laudan. *British Journal for the Philosophy of Science*, **40** (3): 376–388.

1989b. Fresnel, Poisson and the 'White Spot': The Role of Successful Prediction in Theory-Acceptance. In *The Uses of Experiment*, D. Gooding, T. Pinch, and S. Schaffer (eds.), Cambridge: Cambridge University Press, 135–157.

1989c. Structural Realism: The Best of Both Worlds? *Dialectica*, **43** (1–2): 99–124.

2002. New Evidence for Old. In *In the Scope of Logic, Methodology and Philosophy of Science*, P. Gardenfors (ed.), Dordrecht: Kluwer, 191–209.

2005. Prediction and the 'Periodic Law': A Rejoinder to Barnes. *Studies in History and Philosophy of Science Part A*, **36** (4): 817–826.

2014. Prediction and Accommodation Revisited. *Studies in History and Philosophy of Science Part A*, **45**: 54–61.

Zahar, E. 1973a. Why Did Einstein's Programme Supersede Lorentz's? (I). *British Journal for the Philosophy of Science*, **24** (2): 95–123.

1973b. Why Did Einstein's Programme Supersede Lorentz's? (II). *British Journal for the Philosophy of Science*, **24** (3): 223–262.

Index

Printed in the United States
By Bookmasters